CAMBRIDGE

General Editors: I
W. G. Price

SEA LOADS ON SHIPS AND OFFSHORE STRUCTURES

SEA LOADS ON SHIPS AND OFFSHORE STRUCTURES

O. M. Faltinsen
Professor, Department of Marine Technology
Norwegian Institute of Technology

CAMBRIDGE
UNIVERSITY PRESS

PUBLISHED BY THE PRESS SYNDICATE OF THE UNIVERSITY OF CAMBRIDGE
The Pitt Building, Trumpington Street, Cambridge CB2 1RP, United Kingdom

CAMBRIDGE UNIVERSITY PRESS
The Edinburgh Building, Cambridge CB2 2RU, UK http://www.cup.cam.ac.uk
40 West 20th Street, New York, NY 10011-4211, USA http://www.cup.org
10 Stamford Road, Oakleigh, Melbourne 3166, Australia

First published 1990
First paperback edition 1993
Reprinted 1995, 1998

A catalogue record for this book is available from the British Library

ISBN 0 521 45870 6 paperback

Transferred to digital printing 1999

CONTENTS

PREFACE

The material in this book has been continuously developed since the author started to teach hydrodynamics of ships and offshore structures at the Norwegian Institute of Technology (NTH) in 1974. During this period offshore oil activity has played an important role in Norwegian society. Interest in ships for transportation has changed during these years. At the moment there is an increasing interest in developing high-speed marine vehicles for transportation of goods and passengers. In the future it is expected that oil and gas exploration will move into areas of deeper water. Examples on future new areas where ocean engineers and naval architects can be of help is fish farming in open sea, recovery of deep-sea minerals, and development of marine energy resources from temperature gradients and waves. In all these areas there is a need to know about sea loads. This is what this book is all about. The book covers applications in a broad area. This includes conventional ships, high-speed marine vehicles, fixed and floating offshore structures. Many of the applications come through exercises.

Part of the material in the book has been taught for the last year siv.ing (MSc) students at the Department of Marine Technology, NTH. It has also been used in graduate courses at the Department of Ocean Engineering, MIT, when the author was a visiting professor there in two periods from 1980 to 1981 and from 1987 to 1988.

A book on wave-induced motions and loads can easily be very mathematical. The author has tried to avoid this. The hope is that engineers with a non-mathematical background can get a good insight into sea loads on ships and offshore structures by reading the book. However, knowledge in calculus including vector analysis and differential equations is necessary to read the book in detail. The reader should also be familiar with basic hydrodynamics of potential and viscous flow.

I was encouraged by Professor J. N. Newman, MIT, to write the book. Being an editor for the book he has also given me much valuable advice. Dr. Svein Skjørdal has spent a lot of time giving detailed comments on different versions of the manuscript. He has also been helpful in seeing the topics from a practical point of view. Many other people should be thanked for their critical review and contributions.

These include: Dr J. M. R. Graham of Imperial College, London; Dr Martin Greenhow of Brunel University, London; Professor Makoto Ohkusu of Kyushu University; Professor Paul Sclavounos of MIT; Professor Finn Gunnar Nielsen of Norsk Hydro; Professor Enok Palm of University of Oslo; Dr John Grue of University of Oslo, Dr Bjørn Sortland of Marintek; Siv.ing. Terje Nedrelid of Marintek, Professor Bjørnar Pettersen of NTH and Professor Dag Myrhaug of NTH. Graduate students who have been particularly helpful have been Seung Il Ahn, Rong Zhao, Geir Løland, Jan Kvålsvold, Knut Streitlien and Jens Bloch Helmers. Rong Zhao has done the calculations presented in several of the figures and Vigdis O. Dahl is responsible for the skilful drawing of many of the figures. Marianne Kjølaas has typed the many versions of the manuscript in an accurate and efficient way.

1 INTRODUCTION

Knowledge about wave induced loads and motions of ships and offshore structures is important both in design and operational studies. The significant wave height (the mean of the highest one-third of the waves) can be larger than 2 m for 60% of the time in hostile areas like the North Sea. Wave heights higher than 30 m can occur. The mean wave period can be from 15 to 20 s in extreme weather situations and it is seldom below 4 s. Environmental loads due to current and wind are also important. Extreme wind velocities of 40 to 45 m s^{-1} have to be used in the design of offshore structures in the North Sea.

Fig. 1.1 shows five examples of offshore structures. Two of them, the jacket type and the gravity platform, penetrate the sea floor. At present, fixed structures have been built for water depths up to about 300 m. Two of the structures, the semi-submersible and the floating production ship, are free-floating. The tension leg platform (TLP) is restrained from oscillating vertically by tethers, which are vertical anchorlines that are tensioned by the platform buoyancy being larger than the platform weight. Both the ship and the semi-submersible are kept in position by a spread mooring system. An alternative would be to use thrusters and a dynamic positioning system. Pipes (risers) are used as connections between equipment on the sea floor and the platform.

Ships serve a large variety of purposes. Examples are transportation of goods and passengers, naval operations, drilling, marine operations, fishing, sport and leisure activities. Fig. 1.2 shows three types of ships: a monohull, a SWATH and a SES. The monohull is exemplified by a LNG (liquid natural gas) carrier with spherical tanks. SWATH stands for small-waterplane-area, twin-hull ship and consists of two fully-submerged hulls that are connected to the above water structure by one or several thin struts. Between the hulls there may be fitted fins or foils as in Fig. 1.2. SES (surface effect ship) is an air-cushion supported high-speed vehicle where the air-cushion is enclosed on the sides by rigid sidewalls and on the bow and stern by compliant seals. By high speed we mean high Froude number (Fn). This is defined as Fn $=$ $U/(Lg)^{\frac{1}{2}}$($U =$ ship speed, $L =$ ship length, $g =$ acceleration of gravity). A ship is considered a high-speed marine vehicle when Fn $> \approx 0.5$. From a

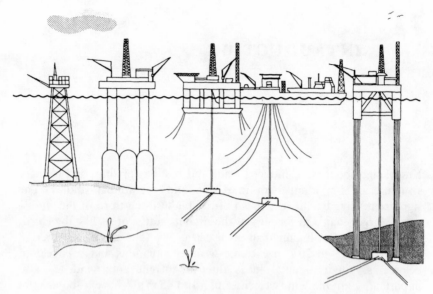

Fig. 1.1. Five types of offshore structures. From left to right we have, jacket, gravity platform, semi-submersible, floating production ship, tension leg platform (TLP). (Partly based on a figure provided by Veritec A/S.)

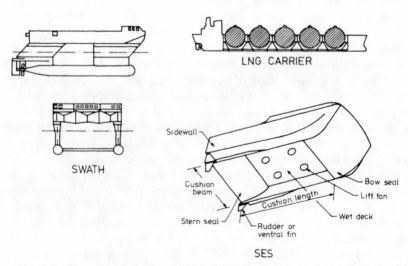

Fig. 1.2. Three types of ships. SWATH (small-waterplane area, twin-hull ship), LNG (liquid natural gas) carrier, SES (surface effect ship).

hydrodynamical view point one can distinguish between ships at zero, normal and high speed. SWATH concepts have been designed for both normal and high-speed applications.

Most of the applications presented in the main text will deal with ships at zero or normal speed and with offshore structures. Applications to high-speed marine vehicles will be given by exercises. We will discuss both wave-induced loads and motions, with motions being the result of integrated hydrodynamic loads on the structure. In the introduction we will give a survey of important wave load and seakeeping problems for ships and offshore structures. Before doing that we need to define the motions.

DEFINITIONS OF MOTIONS

Motions of floating structures can be divided into wave-frequency motion, high-frequency motion, slow-drift motion and mean drift. The oscillatory rigid-body translatory motions are referred to as surge, sway and heave, with heave being the vertical motion (see Fig. 1.3). The oscillatory angular motions are referred to as roll, pitch and yaw, with yaw being rotation about a vertical axis. For a ship, surge is the longitudinal motion and roll is the angular motion about the longitudinal axis.

The wave-frequency motion is mainly linearly-excited motion in the wave-frequency range of significant wave energy. High-frequency mo-

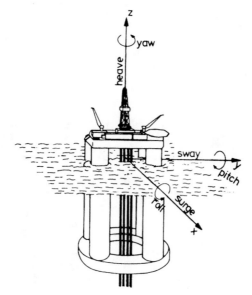

Fig. 1.3. Definition of rigid-body motion modes. Exemplified for a deep concrete floater.

Table 1.1. *Resonant heave oscillations of ships, offshore structures and high speed vehicles*

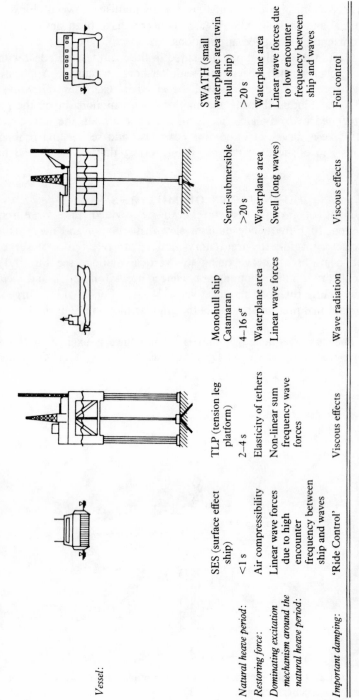

Vessel:	SES (surface effect ship)	TLP (tension leg platform)	Monohull ship Catamaran	Semi-submersible	SWATH (small waterplane area twin hull ship)
Natural heave period:	<1 s	2–4 s	4–16 s[a]	>20 s	>20 s
Restoring force:	Air compressibility	Elasticity of tethers	Waterplane area	Waterplane area	Waterplane area
Dominating excitation mechanism around the natural heave period:	Linear wave forces due to high encounter frequency between ship and waves	Non-linear sum frequency wave forces	Linear wave forces	Swell (long waves)	Linear wave forces due to low encounter frequency between ship and waves
Important damping:	'Ride Control'	Viscous effects	Wave radiation	Viscous effects	Foil control

[a] Rough estimate: $\sqrt{(L/1.5)}$, where L is ship length in metres.

tion is significant for TLPs and is often referred to as 'ringing' and 'springing' and is due to resonance oscillations in heave, pitch and roll of the platform. The restoring forces for the TLP are due to tethers and the mass forces due to the platform. The natural periods of these motion modes are typically 2–4 s which are less than most wave periods. They are excited by non-linear wave effects. 'Ringing' is associated with transient effects and 'springing' is steady-state oscillations.

Similar non-linear effects cause slow drift and mean motions in waves and current. Wind will also induce slow drift and mean motion. Slow drift motion arises from resonance oscillations. For a moored structure it occurs in surge, sway and yaw. The restoring forces are due to the mooring system and the mass forces due to the structure. Typical resonance periods are of the order of 1 to 2 minutes for conventionally moored systems.

Heave is an important response variable for many structures. Table 1.1 illustrates the range of the natural heave periods of different types of marine structures. These include SES, TLPs, monohull ships, catamarans, SWATH ships and semi-submersibles. The table indicates how the natural heave oscillations can be excited. For instance for the SES-hull it occurs due to high encounter frequency between the ship and the waves, while for the SWATH it occurs due to low encounter frequency between the ship and the waves. The table also shows what types of restoring forces can cause heave resonance. For the SES it is the compressibility effect of the air in the cushion. For the monohull ship, catamaran, SWATH and semi-submersible it is due to change in buoyancy forces. This is related directly to the waterplane area of the vessels. Finally we see in Table 1.1 either the most important physical source of natural heave damping or how one artificially increases the damping by control systems.

For the SES it is the heave accelerations and not the heave motions that are important. If no 'ride control' is used, acceleration values of $1.5g$ can occur in relatively calm sea. If the natural heave period is 0.5 s, it means the heave amplitude is ≈ 0.1 m.

A semi-submersible is designed to avoid resonance heave motion and the maximum heave motion in severe sea states will be less than half the maximum wave amplitude.

TRADITIONAL SHIP PROBLEMS

Examples of important seakeeping and wave load problems for ships are illustrated in Fig. 1.4. In particular, vertical accelerations and relative vertical motions between the ship and the waves are important responses. Accelerations determine loads on cargo and equipment and are an important reason for seasickness. The relative vertical motions can be used to evaluate the possibility and damage due to slamming and water on deck. (Slamming means impact between the ship and the water.) For a ship it is important to avoid slamming as well as water on deck because of the resulting local damage of the structures.

Rolling may be a problem from an operational point of view of fishing vessels, crane vessels, passenger ships and naval vessels. Means to reduce

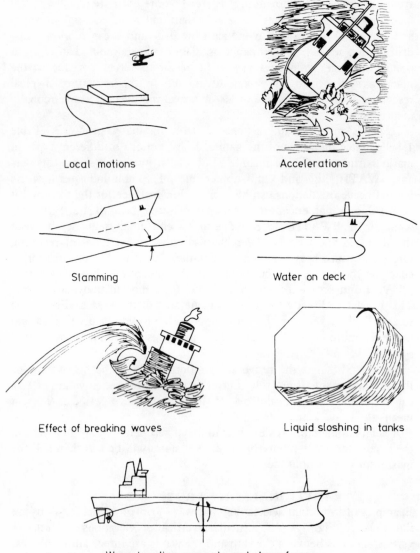

Fig. 1.4. Examples of important seakeeping and wave load problems for ships.

the rolling of a ship are therefore of interest. Examples are bilge keels, anti-roll tanks and active fins. For smaller ships, rolling in combination with either wind, water on deck or motion of the cargo can cause the ship to capsize. Another important reason for capsizing of smaller ships is breaking waves. Several accidents off the Norwegian coast have been explained by breaking waves. Following sea can cause different critical capsizing situations. If the wave profile is stationary relative to the ship, the ship may be statically unstable in roll relative to the waterline defined by the wave profile. The ship may also lose its directional stability in following waves. This can happen when the frequency of encounter between the ship and the waves is small. The result is an altered course relative to the waves. This situation is called 'broaching' and is most critical with respect to capsizing of ships with small static stability.

Liquid sloshing in tanks may be a problem for bulkships, combination ships oil–bulk–ore (OBO), liquid natural gas (LNG) carriers and tankers loading at offshore terminals. There are two reasons why the fluid motion in a tank can be violent. One is that a natural period for the fluid motion in the tank is in a period domain where there is significant ship motion. The other reason is that there is often little damping connected with fluid motion in a tank. If the excitation period is close to a natural period for the tank motion, a strong amplification of the fluid motion in a tank will occur. Liquid sloshing can cause high local pressures as well as large total forces. Both effects may be important in design.

For larger ships, wave-induced bending moments, shear forces and torsional moments are important. More specific problems are whipping and springing. Whipping is transient elastic vibration of the ship hull girder caused for instance by slamming. Springing is steady-state elastic vibration caused by the waves and is of special importantce for larger oceangoing ships and Great Lake carriers. Springing is due to both linear and non-linear excitation mechanisms. The linear exciting forces are associated with waves of small wavelengths relative to the ship length.

Ship motions and sea loads can influence the ship speed significantly due to voluntary and involuntary speed reduction. Voluntary speed reduction means that the ship master reduces the speed due to heavy slamming, water on deck or large accelerations. Involuntary speed reduction is the result of added resistance of the ship due to waves and wind and changes in the propeller efficiency due to waves. The importance of involuntary speed reduction is exemplified in Fig. 1.5. It shows the results of computer calculations for a container ship at a given sea state. The significant wave height $H_{\frac{1}{3}}$ is 8.25 m. The waves are assumed longcrested with different propagation directions relative to the ship. The ship has a length of 185 m. The actual speed at constant engine power is given for different wave headings together with the design speed in still water at the same engine power. For instance in head seas the ship speed is 8 knots $(4.1 \, \text{m s}^{-1})$ compared to 16.2 knots $(8.3 \, \text{m s}^{-1})$ in still water. Depending on the wave direction, the actual ship speed may be lower than that shown in Fig. 1.5. This is due to voluntary speed reduction. Information like this may be used to choose optimum ship routes based on relevant criteria like the lowest fuel consumption or the shortest time of voyage.

Criteria for acceptable levels of ship motions have been discussed in the Nordic co-operative project 'Seakeeping performance of ships' (NORDFORSK, 1987). Considerations have been given to hull safety, operation of equipment, cargo safety, personnel safety and efficiency. General operability limiting criteria for ships are given in Table 1.2. Criteria with regard to accelerations and roll for special types of work and for passenger comfort are given in Table 1.3. The limiting criteria for fast small craft are only indicative of trends. A fast small craft is defined as a vessel under about 35 metres in length with speed in excess of 30 knots. A reason why the vertical acceleration level for fast small craft is set higher than for merchant ships and naval vessels, is that personnel can tolerate higher vertical acceleration when the frequency of oscillation is high.

OFFSHORE STRUCTURE PROBLEMS

For drilling operations heave motion is a limiting factor. The reason is that the vertical motion of the risers has to be compensated and there are limits to how much the motion can be compensated. An example of a

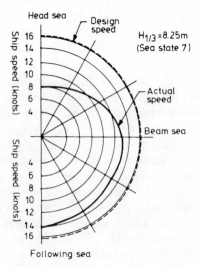

Fig. 1.5. Effect of added ship resistance due to waves and wind (involuntary speed reduction). Ship length = 185 m. ($H_{\frac{1}{3}}$ = significant wave height).

Table 1.2. *General operability limiting criteria for ships* (*NORDFORSK*, 1987)

	Merchant ships	Naval vessels	Fast small craft
Vertical acceleration at forward perpendicular (RMS-value)	$0.275g$ ($L \leqslant 100$ m) $0.05g$ ($L \geqslant 330$ m)[a]	$0.275g$	$0.65g$
Vertical acceleration at bridge (RMS-value)	$0.15g$	$0.2g$	$0.275g$
Lateral acceleration at bridge (RMS-value)	$0.12g$	$0.1g$	$0.1g$
Roll (RMS-value)	6.0 deg	4.0 deg	4.0 deg
Slamming criteria (Probability)	0.03 ($L \leqslant 100$ m) 0.01 ($L \geqslant 300$ m)[b]	0.03	0.03
Deck wetness criteria (Probability)	0.05	0.05	0.05

[a] The limiting criterion for lengths between 100 and 330 m varies almost linearly between the values $L = 100$ m and 330 m, where L is the length of the ship.
[b] The limiting criterion for lengths between 100 and 300 m varies linearly between the values $L = 100$ m and 300 m.

Table 1.3. *Criteria with regard to accelerations and roll* (*NORDFORSK*, 1987)

Root mean square criterion

Vertical acceleration	Lateral acceleration	Roll	Description
$0.20g$	$0.10g$	$6.0°$	Light manual work
$0.15g$	$0.07g$	$4.0°$	Heavy manual work
$0.10g$	$0.05g$	$3.0°$	Intellectual work
$0.05g$	$0.04g$	$2.5°$	Transit passengers
$0.02g$	$0.03g$	$2.0°$	Cruise liner

heave motion criterion is that the heave amplitude should be less than 4 m. It is therefore important to design structures with low heave motion so that it is possible to drill in as high a percentage of the time as possible. Semi-submersibles are examples of structures with very low heave motion in the actual frequency domain. Rolling may also be an important motion mode to evaluate, for example for operation of crane vessels or for transportation of jackets and semi-submersibles on ships and barges. Rolling, pitching and accelerations may represent limiting factors for the operation of process equipment on board a floating production platform.

In the design of mooring systems for offshore structures loads due to current, wind, wave-drift forces and wind- and wave-induced motion are generally of equal importance. There are two important design parameters. One is the breaking strength of the mooring lines. The other is the flexibility of the riser system which means, in practice, for a rigid riser system that the extreme horizontal offsets of the platform relative to the connection point of the riser to the sea floor should be less than say 10% of the water depth.

Wind, current, mean wave drift forces and slowly varying wave drift forces are also important in the design of thrusters and in station keeping of crane vessels, diving vessels, supply ships, offshore loading tankers and pipelaying vessels. Interaction of thrusters with other thrusters, the free-surface and structures may also be important for dynamic positioning systems, towing and marine operations in waves.

Examples of the main objectives of the hydrodynamic analysis of a tension leg platform are, to calculate the vertical dynamic loads on the platform with the purpose of estimating axial forces in the tethers and to calculate the wave elevation in order to evaluate the air gap between the waves and the underside of the platform. The minimum air gap is also an important consideration for other types of platforms.

HYDRODYNAMIC CLASSIFICATION OF STRUCTURES

Both viscous effects and potential flow effects may be important in determining the wave-induced motions and loads on marine structures. Included in the potential flow is the wave diffraction and radiation around the structure. In order to judge when viscous effects or different types of potential flow effects are important, it is useful to refer to a simple picture like Fig. 1.6. This drawing is based on results for horizontal wave forces on a vertical circular cylinder standing on the sea floor and penetrating the free surface. The incident waves are regular. H is the wave height and λ is the wavelength of the incident waves. D is the cylinder diameter. The results are based on the use of Morison's equation (see chapter 7) with a mass coefficient of 2 and a drag coefficient of 1. The linear McCamy & Fuchs (1954) theory has been used in the wave diffraction regime.

Let us try to use the figure for offshore structures. We will consider a regular wave of wave height 30 m and wavelength 300 m. This corresponds to an extreme wave condition. Let us consider wave loads on the caisson of a gravity platform where typical cross-sectional dimensions are 100 m. This implies equivalent H/D- and λ/D-values of 0.3 and 3, respectively. This means that wave diffraction is most important. If we consider the columns of a semi-submersible, a relevant diameter would be approximately 10 m. This implies $\lambda/D = 30$, $H/D = 3$, which means that the hydrodynamic forces are mainly potential flow forces in phase with the undisturbed local fluid acceleration. Wave diffraction and viscous forces are of less significance.

For the legs of a jacket a relevant diameter is approximately 1 m. This implies that viscous forces are most important. By viscous forces we do not mean shear forces, but pressure forces due to separated flow. The examples above are for an extreme wave condition. In an operational wave condition the relative importance of viscous and potential flow effects are different. We should bear in mind that Fig. 1.6 only provides a very rough classification. For instance, resulting forces may be small due to the cancelling out of effects of loads from different parts of the structure.

ENGINEERING TOOLS
Both numerical calculations, model tests and full-scale trials are used to assess wave-induced motions and loads. From an ideal point of view full-scale tests are desirable but expensive and difficult to perform under controlled conditions. It may also be unrealistic to wait for the extreme weather situations to occur. Model tests are therefore needed. A drawback with model tests is the difficulty of scaling test results to full scale results when viscous hydrodynamic forces matter. The geometrical dimensions and equipment of the model test facilities may also limit the experimental possibilities.

Due to the rapid development of computers with large memory capacity and high computational speed, numerical calculations have

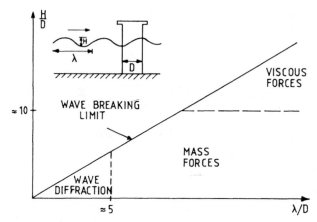

Fig. 1.6. Relative importance of mass, viscous drag and diffraction forces on marine structures.

played an increasingly important role in calculating wave-induced motions and loads on ships and offshore structures. A significant step in the development started about 1970. For offshore structures it was partly connected with the beginning of offshore oil and gas production and exploration in the North Sea. However, it is important to stress that numerical computer programs are also dependent on the development of hydrodynamic theories. More theoretical research is still needed, in particular to increase the knowledge on separated viscous flow and extreme wave effects on ships and offshore structures.

It is unrealistic to expect that computer programs will totally replace model tests in the foreseeable future. The ideal way is to combine model tests and numerical calculations. In some cases computer programs are not reliable. Model tests often give more confidence than computer programs when totally new concepts are tested out.

When computer programs have been validated and the theoretical basis of the computer program has been satisfactorily compared with experimental results, computer programs offer an advantage relative to model tests. Computer programs can often be used in a more efficient way than model tests to evaluate different designs in a large variety of sea conditions. However, sound judgement of results is always important. A basis for this is physical understanding and practical feeling.

One aim of the book is to provide physical understanding to the reader and try to simplify the problems mathematically. In this way one can develop simple tools to evaluate results from model tests, full-scale trials or computer programs.

2 SEA ENVIRONMENT

The intention of this chapter is to provide the basic information on waves, wind and current that is needed to evaluate sea loads and motions acting on ships and offshore structures. It is assumed that the reader has a basic knowledge of fluid mechanics and is familiar with the concepts of velocity potential and Bernoulli's equation. A brief survey of the general aspects of free-surface fluid flow problems based on potential theory is given below.

BASIC ASSUMPTIONS

The sea water is assumed incompressible and inviscid. The fluid motion is irrotational. A velocity potential ϕ can be used to describe the fluid velocity vector $\mathbf{V}(x, y, z, t) = (u, v, w)$ at time t at the point $\mathbf{x} = (x, y, z)$ in a Cartesian coordinate system fixed in space. This means that

$$\mathbf{V} = \nabla \phi \equiv \mathbf{i} \frac{\partial \phi}{\partial x} + \mathbf{j} \frac{\partial \phi}{\partial y} + \mathbf{k} \frac{\partial \phi}{\partial z} \tag{2.1}$$

where $\mathbf{i}$, $\mathbf{j}$ and $\mathbf{k}$ are unit vectors along the x-, y- and z-axes, respectively. A velocity potential has no physical meaning itself, but is introduced because it is convenient in the mathematical analysis of irrotational fluid motion. The fluid is irrotational when the vorticity vector

$$\boldsymbol{\omega} = \nabla \times \mathbf{V} \tag{2.2}$$

is zero everywhere in the fluid. Also, since water is incompressible, i.e. $\nabla \cdot \mathbf{V} = 0$, it follows that the velocity potential has to satisfy the Laplace equation

$$\frac{\partial^2 \phi}{\partial x^2} + \frac{\partial^2 \phi}{\partial y^2} + \frac{\partial^2 \phi}{\partial z^2} = 0 \tag{2.3}$$

The complete mathematical problem of finding a velocity potential of irrotational, incompressible fluid motion consists of the solution of the Laplace equation with relevant boundary conditions on the fluid. We will show examples of boundary conditions later.

The pressure p follows from Bernoulli's equation. If we assume the

z-axis to be vertical and positive upwards we can write

$$p + \rho g z + \rho \frac{\partial \phi}{\partial t} + \frac{\rho}{2} \mathbf{V} \cdot \mathbf{V} = C \tag{2.4}$$

where C is an arbitrary function of time. We will include the time dependence of C in the velocity potential and let C be a constant. Equation (2.4) is true for unsteady, irrotational and inviscid fluid motion. It is assumed that the only external force field is gravity. We will let $z = 0$ correspond to the mean free-surface level. The constant C can be related to the atmospheric pressure or the ambient pressure, as in equation (2.11).

Kinematic boundary conditions
For a fixed body in a moving fluid we have the body boundary condition

$$\frac{\partial \phi}{\partial n} = 0 \quad \text{on the body surface} \tag{2.5}$$

Here $\partial / \partial n$ denotes differentiation along the normal to the body surface. We will define the positive normal direction to be into the fluid domain. Equation (2.5) expresses impermeability, i.e. that no fluid enters or leaves the body surface. The tangential velocity component on a body surface in a potential flow problem is unspecified. If the body is moving with velocity $\mathbf{U}$, equation (2.5) can be generalized to

$$\frac{\partial \phi}{\partial n} = \mathbf{U} \cdot \mathbf{n} \quad \text{on the body surface} \tag{2.6}$$

Here $\mathbf{U}$ can be any type of body velocity. For a rigid body it includes translatory and rotary motion effects in general. This means $\mathbf{U}$ may be different for different points on the body surface.

Before we formulate the kinematic free-surface condition we recall the meaning of the substantial derivative DF/Dt of a function $F(x, y, z, t)$. This expresses the rate of change with time of the function F if we follow a fluid particle in space. Mathematically we can express it as

$$\frac{DF}{Dt} = \frac{\partial F}{\partial t} + \mathbf{V} \cdot \nabla F \tag{2.7}$$

where $\mathbf{V}$ is the fluid velocity at the point (x, y, z) at time t.

As an example let us define the free-surface by the equation

$$z = \zeta(x, y, t) \tag{2.8}$$

where ζ is the wave elevation. We then define the function

$$F(x, y, z, t) = z - \zeta(x, y, t) = 0 \tag{2.9}$$

A fluid particle on the free-surface is assumed to stay on the free-surface. This means it always satisfies equation (2.9) and that $DF/Dt = 0$. The following kinematic boundary condition then applies on the free-surface

$$\frac{\partial}{\partial t}(z - \zeta(x, y, t)) + \nabla\phi \cdot \nabla(z - \zeta(x, y, t)) = 0$$

i.e.

$$\frac{\partial \zeta}{\partial t} + \frac{\partial \phi}{\partial x}\frac{\partial \zeta}{\partial x} + \frac{\partial \phi}{\partial y}\frac{\partial \zeta}{\partial y} - \frac{\partial \phi}{\partial z} = 0 \quad \text{on} \quad z = \zeta(x, y, t) \qquad (2.10)$$

We have expressed here the fluid velocity $\mathbf{V}$ in equation (2.7) by the velocity potential ϕ (see equation (2.1)).

Dynamic free-surface condition

The dynamic free-surface condition is simply that the water pressure is equal to the constant atmospheric pressure p_0 on the free-surface. If we choose the constant C in equation (2.4) as p_0/ρ so that the equation holds with no fluid motion, then

$$g\zeta + \frac{\partial \phi}{\partial t} + \frac{1}{2}\left(\left(\frac{\partial \phi}{\partial x}\right)^2 + \left(\frac{\partial \phi}{\partial y}\right)^2 + \left(\frac{\partial \phi}{\partial z}\right)^2\right) = 0$$

$$\text{on} \quad z = \zeta(x, y, t) \quad (2.11)$$

The free-surface conditions (2.10) and (2.11) are non-linear. We do not know where the free-surface is before we have solved the problem. However, by linearizing the free-surface conditions we are able to simplify the problem and still get sufficient information in most cases. In the study of interactions between linear waves and linear wave-induced motions and loads on ships and offshore structures, the linear free-surface condition will depend on the forward speed or the presence of a current. We assume here that the structure has no forward speed and that the current is zero. Linear theory means that the velocity potential is proportional to the wave amplitude. It is valid if the wave amplitude is small relative to a characteristic wavelength and body dimension. By a Taylor expansion we can transfer the free-surface conditions from the free-surface position $z = \zeta(x, y, t)$ to the mean free-surface at $z = 0$. By keeping linear terms in the wave amplitude we find from equations (2.10) and (2.11) that

$$\frac{\partial \zeta}{\partial t} = \frac{\partial \phi}{\partial z} \quad \text{on} \quad z = 0 \quad \text{(kinematic condition)} \qquad (2.12)$$

$$g\zeta + \frac{\partial \phi}{\partial t} = 0 \quad \text{on} \quad z = 0 \quad \text{(dynamic condition)} \qquad (2.13)$$

Table 2.1. *Velocity potential, dispersion relation, wave profile, pressure, velocity and acceleration for regular sinusoidal propagating waves on finite and infinite water depth according to linear theory*

	Finite water depth	Infinite water depth
Velocity potential	$\phi = \dfrac{g\zeta_a}{\omega}\dfrac{\cosh k(z+h)}{\cosh kh}\cos(\omega t - kx)$	$\phi = \dfrac{g\zeta_a}{\omega}e^{kz}\cos(\omega t - kx)$
Connection between wave number k and circular frequency ω	$\dfrac{\omega^2}{g} = k\tanh kh$	$\dfrac{\omega^2}{g} = k$
Connection between wavelength λ and wave period T	$\lambda = \dfrac{g}{2\pi}T^2\tanh\dfrac{2\pi}{\lambda}h$	$\lambda = \dfrac{g}{2\pi}T^2$
Wave profile	$\zeta = \zeta_a\sin(\omega t - kx)$	$\zeta = \zeta_a\sin(\omega t - kx)$
Dynamic pressure	$p_D = \rho g\zeta_a\dfrac{\cosh k(z+h)}{\cosh kh}\sin(\omega t - kx)$	$p_D = \rho g\zeta_a e^{kz}\sin(\omega t - kx)$
x-component of velocity	$u = \omega\zeta_a\dfrac{\cosh k(z+h)}{\sinh kh}\sin(\omega t - kx)$	$u = \omega\zeta_a e^{kz}\sin(\omega t - kx)$
z-component of velocity	$w = \omega\zeta_a\dfrac{\sinh k(z+h)}{\sinh kh}\cos(\omega t - kx)$	$w = \omega\zeta_a e^{kz}\cos(\omega t - kx)$
x-component of acceleration	$a_1 = \omega^2\zeta_a\dfrac{\cosh k(z+h)}{\sinh kh}\cos(\omega t - kx)$	$a_1 = \omega^2\zeta_a e^{kz}\cos(\omega t - kx)$
z-component of acceleration	$a_3 = -\omega^2\zeta_a\dfrac{\sinh k(z+h)}{\sinh kh}\sin(\omega t - kx)$	$a_3 = -\omega^2\zeta_a e^{kz}\sin(\omega t - kx)$

$\omega = 2\pi/T$, $k = 2\pi/\lambda$, T = Wave period, λ = Wavelength, ζ_a = Wave amplitude, g = Acceleration of gravity, t = Time variable, x = direction of wave propagation, z = vertical coordinate, z positive upwards, $z = 0$ mean waterlevel, h = average waterdepth. Total pressure in the fluid: $p_D - \rho gz + p_0$ (p_0 = atmospheric pressure).

We should note that the free-surface elevation ζ can be found from equation (2.13) when the velocity potential ϕ is known.

Equations (2.12) and (2.13) can be combined to give

$$\frac{\partial^2 \phi}{\partial t^2} + g \frac{\partial \phi}{\partial z} = 0 \quad \text{on} \quad z = 0 \tag{2.14}$$

When the velocity potential ϕ is oscillating harmonically in time with circular frequency ω we can write equation (2.14) as

$$-\omega^2 \phi + g \frac{\partial \phi}{\partial z} = 0 \quad \text{on} \quad z = 0 \tag{2.15}$$

REGULAR WAVE THEORY

By assuming a horizontal sea bottom and a free-surface of infinite horizontal extent we can derive linear wave theory (sometimes called Airy theory) for propagating waves. The free-surface condition (2.15) is then used together with the Laplace equation (2.3) and the sea bottom condition

$$\frac{\partial \phi}{\partial z} = 0 \quad \text{on} \quad z = -h \tag{2.16}$$

where h is the mean water depth.

The derivation of linear wave theory for propagating waves can be found in many textbooks in fluid mechanics (for instance Newman, 1977, chapter 6). Table 2.1 presents the results for both finite and infinite water depths.

We will show how we can derive the results for infinite water depth. The derivation for finite water depth is similar. We assume that the velocity potential can be represented as a product of functions each of which depend on just one independent variable. This means we use the method of 'separation of variables' to solve the Laplace equation. The following solution will satisfy the Laplace equation:

$$\phi = e^{kz}(A \cos kx + B \sin kx) \cos(\omega t + \alpha) \tag{2.17}$$

The quantities A, B and α are arbitrary constants. There also exist other solutions that are proportional to e^{-kz}, $\cos kz$ and $\sin kz$. However, we must disregard them since there should not be any fluid disturbance far down in the fluid, i.e. when $z \to -\infty$. From the free-surface condition (2.15) we can find that there exists a connection between the wave number k and the circular frequency ω used in equation (2.17). We find the dispersion relation:

$$\frac{\omega^2}{g} = k \tag{2.18}$$

In general, equation (2.17) does not represent travelling (propagating) waves. We must then combine solutions like (2.17) so that the x- and t-dependence is like $\cos(\omega t \pm kx + \gamma)$ where γ is a constant phase angle. The plus sign corresponds to waves propagating along the negative x-axis and the minus sign corresponds to waves propagating along the positive x-axis. The velocity of the wave form, i.e. the phase velocity c, is ω/k. This is different from the fluid velocity, which can be found from equation (2.1). The group velocity C_g or the energy propagation velocity is a third type of velocity used in describing wave features. It can be found by

$$C_g = \mathrm{d}\omega/\mathrm{d}k \qquad (2.19)$$

(see e.g. Newman, 1977: pp. 257–66). If we use equation (2.18) we find that C_g is $0.5g/\omega$ for deep water waves, which is half the phase velocity. The group velocity is important when we want to know the propagation velocity of the front of a harmonically oscillating wavetrain. As an example we can consider a wave maker in a model basin that generates harmonically oscillating waves. If we want to know the time it takes for the wave front to reach the 'beach' at the end of the model basin, we should use the group veloctiy (or energy propagation velocity) in the analysis.

We have only discussed waves propagating along the positive or negative x-axis. We can obtain expressions for waves propagating in an arbitrary direction β relative to the x-axis by simply writing the x-, y- and t-dependence of the velocity potential as $\cos(\omega t - kx \cos \beta - ky \sin \beta + \gamma)$. The z-dependence is e^{kz}. We can check this by substituting the expression into the Laplace equation. Another way to see the x-, y-dependence is to first introduce a coordinate system (x', y', z) where there is an angle β between the x-axis and the x'-axis. This means $x = x' \cos \beta + y' \sin \beta$. By substituting this into the expressions for waves propagating along the positive x-axis, and afterwards renaming the (x', y')-coordinates (x, y), we have obtained expressions for waves propagating in an arbitrary direction β. We should note that the waves we are considering have infinitely long crest lengths and that the wave amplitude does not depend on the position along the mean free-surface. We later discuss other type of propagating water waves where this is not true.

We will now discuss the results given in Table 2.1. From the expressions in the table we note that the maximum values of the different physical variables do not happen at the same time. To get a better idea of this we have drawn a picture of the wave elevation at one time instant (see Fig. 2.1). Beneath we have drawn a picture of how the dynamic pressure due to the wave will look. With dynamic pressure we mean the pressure part $-\rho \, \partial \phi/\partial t$ (see equation (2.4)). We have not stated the magnitude of the pressure, which is depth dependent. However, we note that under a wave trough we will get a negative dynamic pressure and under the wave crest there will be a positive dynamic pressure. This is what we might expect from quasi-static considerations. We note further that under the wave crest the fluid velocity is in the wave propagation direction. Beneath a wave trough the fluid velocity is opposite to the wave propagation direction. We should note that the maximum absolute value of the horizontal component of the fluid acceleration is beneath a wave node. The maximum absolute value of the horizontal component of the velocity is beneath either a wave crest or a wave trough.

It should be noted that the linear theory assumes the velocity potential and fluid velocity to be constant from the mean free-surface to the

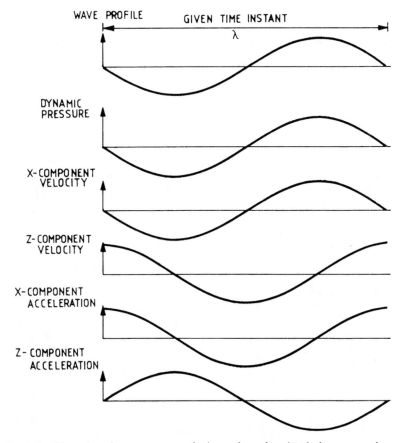

Fig. 2.1. Wave elevation, pressure, velocity and acceleration in long-crested sinusoidal waves propagating along the positive x-axis (see Table 2.1).

free-surface level. This was assumed when the free-surface conditions were formulated. The horizontal velocity distribution shown in Fig. 2.2 for the flow under a wave crest is consistent with linear theory. Fig. 2.2 also shows the velocity under a wave trough, where we have used the analytical velocity distribution up to the free-surface level. It is then implicitly assumed that the difference between the horizontal velocity at the wave trough and the analytical fictitious velocity at $z = 0$ is small compared with the velocity itself. Fig. 2.3 shows how the pressure varies with depth both under a wave crest and a wave trough. It should be noted that the 'hydrostatic' pressure '$-\rho gz$' should cancel the dynamic

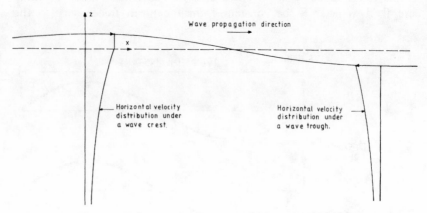

Fig. 2.2. Horizontal velocity distribution under a wave crest and a wave trough according to linear wave theory. (The x- and z-axis have different scales).

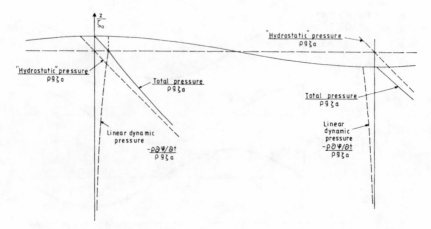

Fig. 2.3. Pressure variation under a wave crest and a wave trough according to linear wave theory.

pressure $-\rho \, \partial\phi/\partial t|_{z=0}$ at the free-surface (see equation (2.13)). This is exactly satisfied at the wave crest in Fig. 2.3 whereas there is a higher-order error under the wave trough. By 'higher-order error' we mean that the error is approximately proportional to $\zeta_a{}^n$, where the order $n \geq 2$. We should note that the dynamic pressure $-\rho \, \partial\phi/\partial t$ half a wavelength down in the fluid is only 4% of its value at $z = 0$. The analytical form of the velocity and pressure distribution that we have used is not the only one that would be consistent with linear theory. For instance, if we assume that the exponential variation of the velocity distribution continues above the mean free-surface, this will also cause an error of $O(\zeta_a{}^2)$ and be consistent with linear theory.

Linear theory represents a first order approximation in satisfying the free surface conditions. It can be improved by introducing higher order terms in a consistent manner – a Stokes' expansion. The next approximation would solve the problem to second order in the parameter ζ_a/λ characterizing the wave amplitude/wavelength ratio of the linear (first-order) solution. Second-order theory means that we keep in a consistent way all terms proportional to $(\zeta_a/\lambda)^2$ and ζ_a/λ. For sinusoidal uni-directional progressive *deep* water waves where the solution in Table 2.1 represents the first-order (linear) solution, it is possible to show that the second-order velocity potential is zero, and that the second-order wave elevation ζ_2 is

$$\zeta_2 = -\tfrac{1}{2}\zeta_a{}^2 k \cos[2(\omega t - kx)] \tag{2.20}$$

By combining this with the first-order solution $\zeta_a \sin(\omega t - kx)$ we see that the second-order solution sharpens the wave crests and makes the troughs more shallow.

The average fluid velocity at one *fixed* point in the fluid is zero according to the second-order theory. However, if we follow fluid particles in time based on linear theory, its average velocity (transport velocity or Stokes drift velocity) over one period will be horizontal and equal to

$$\zeta_a{}^2 \omega k e^{2kz_0} \tag{2.21}$$

for deep water. Here z_0 is the z-coordinate of the fluid particle in its equilibrium position (i.e. when there are no waves). As an example a wave of 100 m wavelength and amplitude ζ_a of 3 m implies a transport velocity at $z_0 = 0$ that is approximately 20% of the orbital velocity at the same place. This has, for instance, consequences for what we mean by current and how current velocities are estimated. In our later discussion on current loads on ships and offshore structures we talk about an ambient constant velocity at *fixed* points. That means we should not include the transport velocity (equation (2.21)) as a current velocity when there are waves present. However, this does not mean that mass transport does not influence the motion of floating marine structures. For a freely floating vessel, especially one moving with the first-order orbital motion, mass transport has an effect on the motions. Ogilvie (1983) has illustrated (see Fig. 2.4) how the transport velocity (equation (2.21)) can be simulated by using the first-order (linear) solution for the fluid velocity. As the fluid particle moves, the local linear velocity will change because the time and the fluid particle position change. The effect over one period is that the fluid particle has moved a horizontal distance equal to the Stokes drift velocity times the period.

Schwartz (1974) has provided a solution in terms of a series expansion that satisfies the exact non-linear free-surface condition within potential theory. This theory can be used to illustrate the limitation of linear theory for regular deep water waves. For waves of steepness $H/\lambda = 0.1$ the exact theory predicts 20% higher maximum wave elevation than the linear approximation. For a wave period 12 s, i.e. a wavelength $\lambda \approx 225$ m, this steepness corresponds to a wave height H of 22.5 m. This is not an unrealistic combination of wave height and period and illustrates, for instance, that linear theory can significantly overpredict the air gap between the waves and a platform deck.

The analytical solution that Schwartz provides assumes periodicity in space. It also assumes that the wave form is symmetric about a vertical axis through the wave crest. This means it cannot be used for studying plunging breakers and irregular steep waves. By plunging breakers we mean waves whose tops are turning over, similar to waves breaking on a beach. Dommermuth et al. (1988) have presented numerical studies of

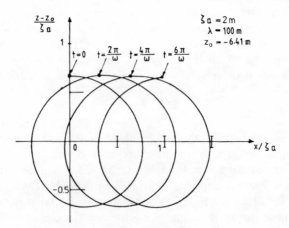

Fig. 2.4. Trajectory of a fluid particle in sinusoidal waves computed from first-order (linear) velocity potential (ζ_a = wave amplitude, λ = wavelength, $z_0 = z$-coordinate of fluid particle at rest.) (Ogilvie, 1983.)

plunging breakers. Good agreement was shown between theory and experiments for a 'two-dimensional' plunging breaker produced by a wave maker, up to the point where the overturning jet re-entered the water surface.

STATISTICAL DESCRIPTION OF WAVES

The present status of computer facilities prevents us from using techniques like that of Dommermuth *et al.* (1988) to obtain statistical estimates for ocean waves, since the estimated CPU-time makes direct calculations unrealistic. In addition there are unsolved physical problems associated with modelling of breaking waves. In practice, linear theory is used to simulate irregular sea and to obtain statistical estimates. The wave elevation of a long-crested irregular sea propagating along the positive x-axis can be written as the sum of a large number of wave components, i.e.

$$\zeta = \sum_{j=1}^{N} A_j \sin(\omega_j t - k_j x + \epsilon_j) \tag{2.22}$$

Here A_j, ω_j, k_j and ϵ_j mean respectively the wave amplitude, circular frequency, wave number and random phase angle of wave component number j. The random phase angles ϵ_j are uniformly distributed between 0 and 2π and constant with time. For deep water waves ω_j and k_j are related by the dispersion relationship (equation (2.18)). The wave amplitude A_j can be expressed by a wave spectrum $S(\omega)$. We can write

$$\tfrac{1}{2}A_j^2 = S(\omega_j)\,\Delta\omega \tag{2.23}$$

where $\Delta\omega$ is a constant difference between successive frequencies. The instantaneous wave elevation is Gaussian distributed with zero mean and variance σ^2 equal to $\int_0^\infty S(\omega)\,d\omega$, which can be shown by using the definition of mean value and variance applied to the 'signal' represented by equation (2.22). We find for instance that $\sigma^2 = \sum_{j=1}^{N} A_j^2/2$. By using equation (2.23) and letting $N \to \infty$ and $\Delta\omega \to 0$, we find that $\sigma^2 = \int_0^\infty S(\omega)\,d\omega$. The relationship between a time domain solution of the waves (i.e. equation (2.22)) and the frequency domain representation of the waves by a wave spectrum $S(\omega)$ is illustrated in Fig. 2.5.

The wave spectrum can be estimated from wave measurements (Kinsman, 1965). It assumes that we can describe the sea as a stationary random process. This means in practice that we are talking about a limited time period in the range from $\tfrac{1}{2}$ hour to maybe 10 hours. In the literature this is often referred to as a short-term description of the sea.

Recommended sea spectra from ISSC (International Ship and Offshore Structures Congress) and ITTC (International Towing Tank Conference) are often used to calculate $S(\omega)$. For instance, for open sea conditions the 15th ITTC recommended the use of ISSC spectral formulation for fully developed sea

$$\frac{S(\omega)}{H_{\frac{1}{3}}^2 T_1} = \frac{0.11}{2\pi} \left(\frac{\omega T_1}{2\pi}\right)^{-5} \exp\left[-0.44\left(\frac{\omega T_1}{2\pi}\right)^{-4}\right] \quad (2.24)$$

where $H_{\frac{1}{3}}$ is the significant wave height defined as the mean of the one third highest waves and T_1 is a mean wave period defined as

$$T_1 = 2\pi m_0/m_1$$

where

$$m_k = \int_0^\infty \omega^k S(\omega)\,\mathrm{d}\omega \quad (2.25)$$

$H_{\frac{1}{3}}$ is often redefined as

$$H_{\frac{1}{3}} = 4\sqrt{m_0} \quad (2.26)$$

which gives a value which is usually close to the $H_{\frac{1}{3}}$ defined above.

Equation (2.24) satisfies equation (2.26). Strictly speaking this relation is only true for a narrow-banded spectrum and when the instantaneous value of the wave elevation is Gaussian distributed.

The spectrum given by equation (2.24) is the same as the modified

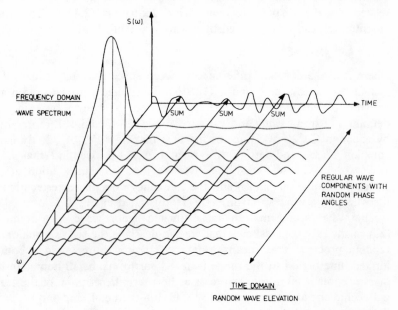

Fig. 2.5. Figure illustrating the connection between a frequency domain and time domain representation of waves in a long-crested short term sea state.

Pierson–Moskowitz spectrum, where it is more usual to use the mean wave period T_2 defined as

$$T_2 = 2\pi(m_0/m_2)^{\frac{1}{2}} \tag{2.27}$$

The following relation exists between T_1 and T_2 for the spectrum given by equation (2.24)

$$T_1 = 1.086T_2 \tag{2.28}$$

The period T_0 corresponding to the peak frequency of the spectrum can for the same spectrum be written as

$$T_0 = 1.408T_2 \tag{2.29}$$

T_0 is also referred to as the modal period.

The spectrum formulation given by equation (2.24) is shown in Fig. 2.6. We note that there is little energy density when $\omega T_2/(2\pi) < 0.5$. For large frequencies the wave spectrum decays like ω^{-5}.

The 17th ITTC recommended the following JONSWAP (Joint North Sea Wave Project) type spectrum for limited fetch:

$$S(\omega) = 155 \frac{H_{\frac{1}{3}}^2}{T_1^4\omega^5} \exp\left(\frac{-944}{T_1^4\omega^4}\right)(3.3)^Y \quad (\text{m}^2\text{s}) \tag{2.30}$$

where

$$Y = \exp\left(-\left(\frac{0.191\omega T_1 - 1}{2^{\frac{1}{2}}\sigma}\right)^2\right)$$

and

$$\sigma = 0.07 \quad \text{for} \quad \omega \leqslant 5.24/T_1$$
$$= 0.09 \quad \text{for} \quad \omega > 5.24/T_1$$

This formulation can be used with the other characteristic periods by the substitution

$$T_1 = 0.834T_0 = 1.073T_2 \tag{2.31}$$

The JONSWAP spectrum is shown in Fig. 2.6. We note that the peak value of the modified Pierson–Moskowitz (ISSC) spectrum occurs at a different $(\omega T_2/2\pi)$-value than the JONSWAP spectrum. This can be seen from equations (2.29) and (2.31).

A good approximation to the probability density function for the maxima (peak values) A of the wave elevation can be obtained from the Rayleigh distribution given by

$$p(A) = \frac{A}{m_0} e^{-A^2/(2m_0)} \tag{2.32}$$

where m_0 is related to $H_{\frac{1}{3}}$ by equation (2.26). Strictly speaking the Rayleigh distribution depends on the wave spectrum being narrow-banded, which is an approximation for the spectra we have discussed. In deriving the Rayleigh distribution, it is also assumed that the instantaneous value of the wave elevation is Gaussian distributed.

We can simulate a seaway by using equation (2.22), but this expression repeats itself after a time $2\pi/\Delta\omega$. A large number N of wave components are therefore needed. A practical way to avoid this is to choose a random frequency in each frequency interval $(\omega_j - \Delta\omega/2,\ \omega_j + \Delta\omega/2)$ and calculate the wave spectrum with those frequencies. The number of wave components ought to be about 1000. This depends partly on the selection of the minimum and maximum frequency component. The minimum

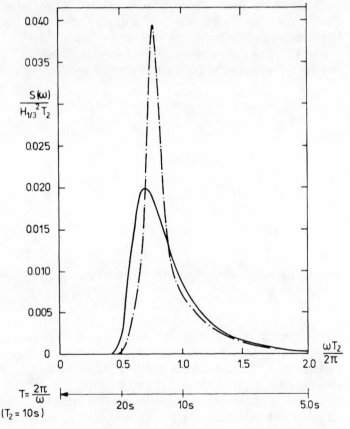

Fig. 2.6. Examples of wave spectra. ($H_{\frac{1}{3}}$ = significant wave height, T_2 = mean wave period). Modified Pierson–Moskowitz spectrum ———— (see equation (2.24)), JONSWAP spectrum ——·——· (see equation (2.30)).

frequency component ω_{min} is easier to select than the maximum frequency component ω_{max}. For instance if a Pierson–Moskowitz spectrum is used $\omega_{min} \approx \pi/T_2$. The wave energy drops off more slowly for larger frequencies than for small frequencies. We should therefore investigate the results for different values of ω_{max}, to ensure that the results do not depend on the selection of ω_{max}. We have only shown how we can simulate the wave elevation ζ. For instance, if we want to simulate the horizontal fluid velocity u and acceleration a_1 we can write

$$u = \sum_{j=1}^{N} \omega_j A_j e^{k_j z} \sin(\omega_j t - k_j x + \epsilon_j) \qquad (2.33)$$

$$a_1 = \sum_{j=1}^{N} \omega_j^2 A_j e^{k_j z} \cos(\omega_j t - k_j x + \epsilon_j) \qquad (2.34)$$

by superposition of the results for regular waves given in Table 2.1. The random phase angles ϵ_j are the same for both ζ, u and a_1.

Fig. 2.7 shows some examples of simulations of the wave elevation. The same sea spectrum and the same duration is used in each simulation. The reason for the differences in the results is the random selection of frequencies and phase angles. We note that the largest amplitude in each simulation (realization) is different. By selecting a large number of realizations we find that the extreme values have their own probability distribution. This is, for instance, discussed by Ochi (1982). In practice the most probable largest value A_{max} is often used. This can be approximated as

$$A_{max} = \left(2m_0 \log \frac{t}{T_2} \right)^{\frac{1}{2}} \qquad (2.35)$$

where t is the time duration and 'log' is the natural logarithm. We should note that A_{max} is the most probable largest value. With that we imply that there is a probability for A_{max} to be exceeded during the time t (Ochi, 1982). The most probable maximum crest-to-trough wave height H_{max} during the same time period is simply $2A_{max}$.

The effect of short-crestedness may be important. A short-crested sea is characterized by a two-dimensional wave spectrum, which in practice is often written as

$$S(\omega, \theta) = S(\omega)f(\theta) \qquad (2.36)$$

where θ is an angle measuring wave propagation direction of elementary wave components in the sea. An example of $f(\theta)$ might be

$$f(\theta) = \begin{cases} \dfrac{2}{\pi} \cos^2 \theta, & -\pi/2 \leqslant \theta \leqslant \pi/2 \\ 0; & \text{elsewhere} \end{cases} \qquad (2.37)$$

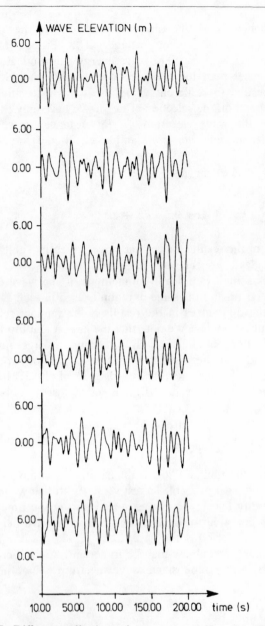

Fig. 2.7. Different realizations of a wave record $H_{\frac{1}{3}} = 8$ m, $T_2 = 10$ s, ISSC-spectrum.

where $\theta = 0$ corresponds to the main wave propagation direction. Other ways of representing a short-crested sea spectrum may be found in the report of the 10th ISSC. For short-crested sea equation (2.22) can be generalized to

$$\zeta = \sum_{j=1}^{N} \sum_{k=1}^{K} (2S(\omega_j, \theta_k)\, \Delta\omega_j\, \Delta\theta_k)^{\frac{1}{2}}$$

$$\times \sin(\omega_j t - k_j x \cos \theta_k - k_j y \sin \theta_k + \epsilon_{jk}) \qquad (2.38)$$

Long-term sea state

So far we have discussed a 'short-term' description of the sea, which means the significant wave height and the mean wave period are assumed constant during the time considered. The significant wave height and mean wave period will vary in a 'long-term' description of the sea. In order to construct a 'long-term' prediction of the sea we need to know the joint frequency of the significant wave height and the mean wave period. An example is presented in Table 2.2. These data are representative for the northern North Sea. The frequency table shows, for instance, that the probability of the significant wave height being between 3 and 4 m and the spectral peak period being 10 s is $2960/100\,001 = 0.0296$. It also shows that the probability of the significant wave height being larger than 2 m is $1 - (8636 + 32\,155)/100\,001 = 0.59$. This table can be used in many ways, for example to obtain long-term statistics of the wave amplitude or wave height. For each significant wave height interval we find the probability of occurrence p_j from the table. For instance the probability that $H_{\frac{1}{3}}$ is between 4 and 5 m is $9118/100\,001$. Since the probability function for the maxima of the wave elevation for given significant wave height follows a Rayleigh distribution (see equation (2.32)) we can obtain the long-term probability as a simple summation, i.e.

$$P(H) = 1 - \sum_{j=1}^{M} e^{(-2H^2/(H_{\frac{1}{3}}^{(j)})^2)} p_j \qquad (2.39)$$

Here $P(H)$ is the long term probability that the wave height does not exceed H. By wave height we mean crest-to-trough wave height and not significant wave height. When using equation (2.32), we have set $H = 2A$. If we use Table 2.2 we see that $M = 15$, $H_{\frac{1}{3}}^{(1)} = 0.5$ m, $H_{\frac{1}{3}}^{(2)} = 1.5$ m and so forth. The probability level $Q = 1 - P(H)$ and the number of response cycles are related by

$$Q = \frac{1}{N}$$

Table 2.2. *Joint frequency of significant wave height and spectral peak period. Representative data for the northern North Sea*

Significant wave height (m) (upper limit of interval)	Spectral peak period (s)																			Sum
	3	4	5	6	7	8	9	10	11	12	13	14	15	16	17	18	19	21	22	
1	59	403	1061	1569	1634	1362	982	643	395	232	132	74	41	22	12	7	4	2	2	8636
2	9	212	1233	3223	5106	5814	5284	4102	2846	1821	1098	634	355	194	105	56	30	16	17	32155
3	0	8	146	831	2295	3896	4707	4456	3531	2452	1543	901	497	263	135	67	33	16	15	25792
4	0	0	6	85	481	1371	2406	2960	2796	2163	1437	849	458	231	110	50	22	10	7	15442
5	0	0	0	4	57	315	898	1564	1879	1696	1228	748	398	191	84	35	13	5	3	9118
6	0	0	0	0	3	39	207	571	950	1069	885	575	309	142	58	21	7	2	1	4839
7	0	0	0	0	0	2	27	136	347	528	533	387	217	98	37	12	4	1	0	2329
8	0	0	0	0	0	0	2	20	88	197	261	226	138	64	23	7	2	0	0	1028
9	0	0	0	0	0	0	0	2	15	54	101	111	78	39	14	4	1	0	0	419
10	0	0	0	0	0	0	0	0	2	11	30	45	39	22	8	2	1	0	0	160
11	0	0	0	0	0	0	0	0	0	2	7	15	16	11	5	1	0	0	0	57
12	0	0	0	0	0	0	0	0	0	0	1	4	6	5	2	1	0	0	0	19
13	0	0	0	0	0	0	0	0	0	0	0	1	2	2	1	0	0	0	0	6
14	0	0	0	0	0	0	0	0	0	0	0	0	0	1	0	0	0	0	0	1
15	0	0	0	0	0	0	0	0	0	0	0	0	0	0	0	0	0	0	0	0
Sum	68	623	2446	5712	9576	12799	14513	14454	12849	10225	7256	4570	2554	1285	594	263	117	52	45	100001

For instance, during 100 years by assuming an average period 7 s we find

$$N = \frac{100 \cdot 365 \cdot 24 \cdot 3600}{7} = 4.5 \cdot 10^8,$$

i.e. $Q = 10^{-8.7}$. By using (2.39) we can find the value H for which $Q = 10^{-8.7}$. We have then found the wave height of what is called the '100 year wave' in offshore engineering.

We can also use Table 2.2 as a basis for obtaining long-term probability of response variables like the heave motion of a ship. In that case we need to combine the joint probability of the significant wave height and mean wave period with the short term statistical distribution of the heave amplitude. Another way we can use Table 2.2 is in operation studies. We can find out the percentage of time during a year that an operation can be performed according to limiting criteria for the operation. An example of such a criterion might be that an offshore loading can only be performed for significant wave heights less than 7 m. Table 2.2 tells us that it is not possible to do the loading for 1.7% of the time during a year. A shortcoming of a joint frequency table of wave periods and significant wave heights is that it does not tell anything about the duration of the sea states. This is important information for studies of marine operations. In addition we need statistical information on sea direction.

WIND

Fig. 2.8 shows the cumulative distribution function for one hour mean wind speed at 10 m above mean sea level (MSL) using representative data for the northern North Sea. The extreme wind speed with return period of 100 years can be found to be 41 m s^{-1}. Time average wind speeds over prescribed time periods are used in calculating steady wind forces on marine structures. However, fluctuating wind forces due to gust may also be of importance. In some cases gust winds can excite resonant oscillations of offshore structures. An example is slow-drift horizontal motion of moored structures. There exist different spectral formulations of wind gust; for example the Harris wind spectrum is given by

$$\frac{f \cdot S(f)}{U_{10}^2} = \frac{4\kappa \bar{f}}{(2 + \bar{f}^2)^{\frac{5}{6}}} \tag{2.40}$$

where U_{10} is the one hour mean wind speed at 10 m above sea level, f is the frequency in Hz and $\bar{f} = Lf/U_{10}$. An example of the scale length L is 1800 m and of the surface drag coefficient κ is 0.0030. Equation (2.40) is not recommended for frequencies lower than 0.01 Hz. A discussion of different wind spectrum formulations is given in the report of the 10th ISSC.

Table 2.3. Annual sea state occurrences in the North Atlantic and North Pacific (Lee et al., 1985)

Sea state no.	Significant wave height (m)		Sustained wind speed (knots)[a]		North Atlantic			North Pacific		
					Percentage probability of sea state	Modal wave period (s)		Percentage probability of sea state	Modal wave period (s)	
	Range	Mean	Range	Mean		Range[b]	Most probable[c]		Range[b]	Most probable[c]
0–1	0–0.1	0.05	0–6	3	0.70	—	—	1.30	—	—
2	0.1–0.5	0.3	7–10	8.5	6.80	3.3–12.8	7.5	6.40	5.1–14.9	6.3
3	0.5–1.25	0.88	11–16	13.5	23.70	5.0–14.8	7.5	15.50	5.3–16.1	7.5
4	1.25–2.5	1.88	17–21	19	27.80	6.1–15.2	8.8	31.60	6.1–17.2	8.8
5	2.5–4	3.25	22–27	24.5	20.64	8.3–15.5	9.7	20.94	7.7–17.8	9.7
6	4–6	5	28–47	37.5	13.15	9.8–16.2	12.4	15.03	10.0–18.7	12.4
7	6–9	7.5	48–55	51.5	6.05	11.8–18.5	15.0	7.00	11.7–19.8	15.0
8	9–14	11.5	56–63	59.5	1.11	14.2–18.6	16.4	1.56	14.5–21.5	16.4
>8	>14	>14	>63	>63	0.05	18.0–23.7	20.0	0.07	16.4–22.5	20.0

[a] Ambient wind sustained at 19.5 m above surface to generate fully-developed seas. To convert to another altitude, H_2, apply $V_2 = V_1(H_2/19.5)^{\frac{1}{7}}$.

[b] Minimum is 5 percentile and maximum is 95 percentile for periods given wave height range.

[c] Based on periods associated with central frequencies included in Hindcast Climatology.

In addition to the information above we need to know how the wind varies with the height above the sea level, in what direction the wind is blowing and the joint probability between waves and wind. We will not go into details about this here. Instead we will present a very simplified picture of corresponding values between significant wave height, modal wave period and sustained wind speed in Table 2.3. This table also gives the percentage probability of sea states. The data are valid for open ocean in the North Atlantic and North Pacific.

CURRENT

State of the art information on ocean currents from a design point of view has been presented by the 10th ISSC. The surface current velocity U is divided into the following components

$$U = U_t + U_w + U_s + U_m + U_{\text{set-up}} + U_d$$

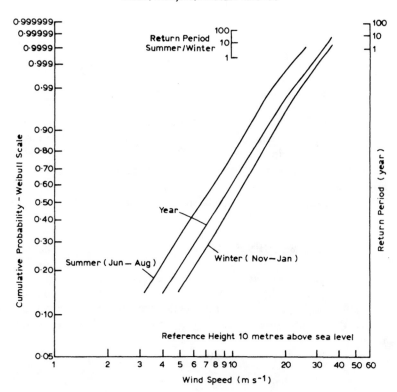

Fig. 2.8. Cumulative distribution function for the one hour mean wind speed at 10 m above MSL representative for the northern North Sea.

where U_t is the tidal component,

U_w is the component generated by local wind,

U_s is the component generated by Stokes drift (see equation (2.21) valid for regular waves)

U_m is the component from major ocean circulation, if any, depending on geographical location (Gulf Stream etc.),

U_{set-up} is the component due to set-up phenomena and storm surges,

U_d is the local density-driven current governed by strong density jumps in the upper ocean.

The depth dependence of the tidal component $U_t(z)$ and the local wind component $U_w(z)$ are given as

$$U_t(z) = \left[\begin{array}{l} U_t(0), \text{ for } -(h - 10) \leqslant z \leqslant 0 \\[2ex] U_t(0) \log_{10}\left(1 + \frac{9z}{10 - h}\right), \text{ for } -h < z < -(h - 10) \end{array} \right.$$

$$U_w(z) = \left[\begin{array}{l} U_w(0) \frac{(h_0 + z)}{h_0}, \text{ for } -h_0 \leqslant z \leqslant 0 \\[2ex] 0, \text{ for } z < -h_0 \end{array} \right.$$

where h is water depth. Distances are in metres and h_0 can be chosen as 50 m. As a first approximation one may set

$$U_w(0) = 0.02 U_{10}$$

where U_{10} is the wind velocity measured 10 m above sea level. However, the Stokes drift is included in this formula. This can be significant (see equation (2.21)) and should not be included if the current velocity is wanted at a *fixed* point in space. The tidal velocity component depends on the location. In open sea it may be up to $\approx 0.5 \text{ m s}^{-1}$. A total current velocity typical for design of offshore structures in the North Sea is 1 m s^{-1}.

EXERCISES

2.1 Resonant fluid motion in a rectangular tank

Consider a rectangular tank partly filled with water. The water depth is constant and equal to h. The tank breadth is $2b$. Assume two-dimensional fluid motion in a (y, z) plane and that the tank is not moving.

(a) Show that the velocity potential

$$\phi_T = A \cosh(k(z + h)) \cos ky \cos \omega t \tag{2.41}$$

satisfies the Laplace equation and the boundary condition on the bottom of the tank.

(b) Which values of k are possible for the velocity potential to satisfy the boundary conditions on the side walls of the tank?

(c) Show from the free-surface condition that the only periods when fluid motion is possible (i.e. natural periods) are given by

$$T_N = 2\pi \left/ \left(\frac{gn\pi}{b} \tanh\left(\frac{n\pi h}{b}\right) \right)^{\frac{1}{2}} \right., n = 1, 2, 3, \ldots \qquad (2.42)$$

Derive an approximate formula when $h/b \to 0$.

(d) Describe the fluid motion at the free-surface as a function of time.

2.2 Propagating water waves

Consider a velocity potential

$$\phi = Ae^{kz}\left(\frac{1}{r}\right)^{\frac{1}{2}} \cos(\omega t - kr) \qquad (2.43)$$

where $r = (y^2 + x^2)^{\frac{1}{2}}$ and A is a constant. Assume deep water and a free surface of infinite horizontal extent.

(a) Is the Laplace equation satisfied everywhere in the fluid?

(b) In what directions are the waves described by equation (2.43) propagating?

(c) How is the wave amplitude varying in space?

2.3 Wave kinematics in regular waves

Consider a model basin with a wave maker in one end that generates long-crested regular waves with circular frequency ω. In the following calculation you can assume a wave period of 2 s and a wave amplitude of 0.25 m. The tank length is 100 m.

(a) Approximately how long will it take for a wave front to propagate from the wave maker to the other end of the tank? Assume the water depth is infinite.

(b) Consider a small cork that is floating on the water and is not disturbing the wave field. Approximately how long will it take for the cork to move from the wave maker to the other end of the tank?

(c) What is the maximum fluid velocity in the tank?

(d) Consider an observer situated along the tank side and a time after the wave front has passed by. How long does it take between two succeeding wave crests to pass the observer? What is the phase of the wave elevation 1.5 m closer to the wave maker relative to the wave elevation at the position of the observer?

(e) What are the answers to the questions in (d) if the observer is either moving with a velocity 1 m s^{-1} toward the wave maker or away from the wave maker?

(f) What would the answers to questions (c), (d) and (e) be if the water depth is either 10 m or 1 m?

2.4 Sea spectrum

Assume a sea spectrum as presented in Fig. 2.9.

(a) Show from the definitions of $H_{\frac{1}{3}}$ and T_1 that $a = 1.5$ and $b = (32\pi)^{-1}$.

(b) What is the relationship between T_2 and T_1?

2.5 Standard deviations of fluid velocity and acceleration in short-term sea states

(a) What are the standard deviations of the horizontal velocity and accelerations given by equations (2.33) and (2.34) when $N \to \infty$ and $\Delta\omega \to 0$.

(b) Assume the wave spectrum is either given in the form of equation (2.24) or equation (2.30). Explain why the standard deviation of the fluid acceleration does not exist for all points in the fluid. How should the wave spectrum look for the standard deviation of the fluid acceleration to exist for all points in the fluid?

(c) Generalize the results in (a) to short-crested sea by using expressions similar to equation (2.38). Discuss the results.

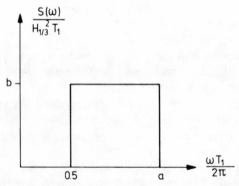

Fig. 2.9. Sea spectrum used in exercise 2.4. ($H_{\frac{1}{3}}$ = significant wave height, T_1 = mean wave period based on first moment of the spectrum).

3 LINEAR WAVE-INDUCED MOTIONS AND LOADS ON FLOATING STRUCTURES

Linear theory can, to a large extent, describe the wave-induced motions and loads on semi-submersibles, ships and other large-volume structures. However, non-linear effects are important in severe sea states and in describing horizontal motions of moored structures.

Consider a structure in incident regular waves of amplitude ζ_a. The wave steepness is small, i.e. the waves are far from breaking. Linear theory means that the wave-induced motion and load amplitudes are linearly proportional to ζ_a.

RESPONSE IN IRREGULAR SEA

A useful consequence of linear theory is that we can obtain results in irregular waves by adding together results from regular waves of different amplitudes, wavelengths and propagation directions.

Let us explain what we mean by considering a long-crested irregular sea described by a sea spectrum $S(\omega)$. We write the wave elevation as

$$\zeta = \sum_{j=1}^{N} A_j \sin(\omega_j t - k_j x + \epsilon_j) \tag{3.1}$$

where

$$\tfrac{1}{2}A_j^2 = S(\omega_j)\,\Delta\omega \tag{3.2}$$

(see equations (2.22) and (2.23) with explanations). Because of linearity we can analyse the response to each wave component in equation (3.1) separately. Examples of response types could be heave or pitch motion of a floating structure. Formally we can write the steady state response as

$$A_j \,|H(\omega_j)|\, \sin(\omega_j t + \delta(\omega_j) + \epsilon_j) \tag{3.3}$$

Here $|H(\omega_j)|$ is called the transfer function, which is the response amplitude per unit wave amplitude. We note also that there is a phase angle $\delta(\omega_j)$ associated with the response. Both $|H(\omega_j)|$ and $\delta(\omega_j)$ are functions of the frequency of oscillation ω_j. The response can be any linear wave-induced motion or load on a structure. Having obtained the response due to one wave component we can linearly superpose the response from the different wave components, i.e. we can write

$$\sum_{j=1}^{N} A_j |H(\omega_j)| \sin(\omega_j t + \delta(\omega_j) + \epsilon_j) \tag{3.4}$$

In the limit as $N \to \infty$ and $\Delta\omega \to 0$ the variance of the response σ_r^2 can be found in the same way as for the waves (see the text after equation (2.23)). We find

$$\sigma_r^2 = \int_0^\infty S(\omega) |H(\omega)|^2 \, d\omega \tag{3.5}$$

The Rayleigh probability function can be used as a good approximation to find the probability density function for the maxima (peak values) of the response R, i.e. we can write

$$p(R) = \frac{R}{\sigma_r^2} e^{-0.5 R^2 / \sigma_r^2} \tag{3.6}$$

Here R may for instance be heave maxima and σ_r the standard deviation of the heave motion.

The most probable largest value $R_{\max}$ during a 'short-term' time t is then

$$R_{\max} = \left(2\sigma_r^2 \log \frac{t}{T_2} \right)^{\frac{1}{2}} \tag{3.7}$$

This is valid for a given value of the significant wave height $H_{\frac{1}{3}}$ and the mean wave period T_2, i.e. for a short-term description of the sea. Strictly speaking we should have used the mean period for the response-variable instead of T_2 in equation (3.7). However, for linear wave-induced motions and loads this will mean a negligible difference in the estimation of $R_{\max}$. By combining the Rayleigh distribution with a joint frequency table for $H_{\frac{1}{3}}$ and the modal wave period (or mean wave period, see Table 2.2) we can obtain the long-term probability in a similar way to equation (2.39). An important difference is that we have to sum over both period and wave height, i.e. we must write

$$P(R) = 1 - \sum_{j=1}^{M} \sum_{k=1}^{K} \exp(-0.5 R^2 / (\sigma_r^{jk})^2) p_{jk} \tag{3.8}$$

where $P(R)$ is the long term probability that the peak value of the response does not exceed R, and σ_r^{jk} is the standard deviation of the response for a mean $H_{\frac{1}{3}}$ and modal period in significant wave height interval j and modal wave period interval k. Further p_{jk} is the joint probability for a significant wave height and a modal wave period to be in interval-numbers j and k, respectively. For instance, by referring to Table 2.2, the joint probability for the modal (spectral peak) period to be

10 s and $H_{\frac{1}{3}}$ to be between 2 and 3 m is 4456/100 001. The values of M and K are respectively 15 and 19. The probability level $Q = 1 - P(R)$ and the number of response cycles N are related by $Q = 1/N$. A return period of 100 years corresponds to $Q = 10^{-8.7}$. The corresponding response amplitude R can then be found from equation (3.8).

RESPONSE IN REGULAR WAVES

Since it is possible to obtain results in irregular seas by linearly superposing results from regular wave components, it is sufficient from a hydrodynamical point of view to analyse a structure in incident regular sinusoidal waves of small wave steepness. This will be done in the following text. We will assume a steady state condition. This means there are no transient effects present due to initial conditions. It implies that the linear dynamic motions and loads on the structure are harmonically oscillating with the same frequency as the wave loads that excite the structure. The hydrodynamic problem in regular waves is normally dealt with as two sub-problems namely:

A: The forces and moments on the body when the structure is restrained from oscillating and there are incident regular waves. The hydrodynamic loads are called *wave excitation loads* and composed of so-called Froude–Kriloff and diffraction forces and moments.

B: The forces and moment on the body when the structure is forced to oscillate with the wave excitation frequency in any rigid-body motion mode. There are no incident waves. The hydrodynamic loads are identified as *added mass, damping* and *restoring* terms.

Due to linearity the forces obtained in A and B can be added to give the total hydrodynamic forces. This is illustrated in Fig. 3.1.

Before we go into detail and describe the different hydrodynamic loads, we will define a coordinate system and the rigid body motion modes. A right-handed coordinate system (x, y, z) fixed with respect to the mean position of the body is used, with positive z vertically upwards through the centre of gravity of the body and the origin in the plane of the undisturbed free-surface. If the body moves with a mean forward speed, the coordinate system moves with the same speed. The body is normally assumed to have the x–z plane as a plane of symmetry. Let the translatory displacements in the x-, y- and z-directions with respect to the origin be η_1, η_2 and η_3 respectively so that η_1 is the surge, η_2 is the sway and η_3 is the heave displacement. Furthermore, let the angular displacement of the rotational motion about the x-, y- and z-axis be η_4, η_5 and η_6 respectively so that η_4 is the roll, η_5 is the pitch and η_6 is the yaw angle. The coordinate system and the translatory and angular displacement conventions are shown for the case of a ship in Fig. 3.2.

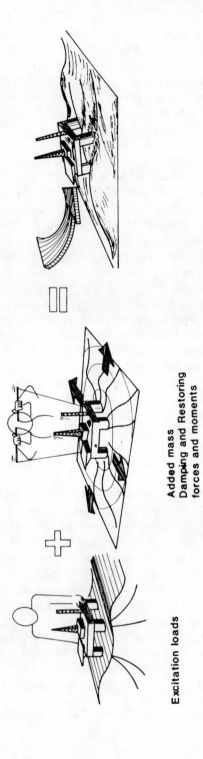

Excitation loads

**Added mass
Damping and Restoring
forces and moments**

Fig. 3.1. Superposition of wave excitation, added mass, damping and restoring loads.

The motion of any point on the body can be written as

$$s = \eta_1 \mathbf{i} + \eta_2 \mathbf{j} + \eta_3 \mathbf{k} + \boldsymbol{\omega} \times \mathbf{r}$$

where '$\times$' denotes vector product and

$$\boldsymbol{\omega} = \eta_4 \mathbf{i} + \eta_5 \mathbf{j} + \eta_6 \mathbf{k}, \qquad \mathbf{r} = x\mathbf{i} + y\mathbf{j} + z\mathbf{k},$$

and $\mathbf{i}$, $\mathbf{j}$, $\mathbf{k}$ are unit vectors along the x-, y- and z-axis, respectively.
This means

$$s = (\eta_1 + z\eta_5 - y\eta_6)\mathbf{i} + (\eta_2 - z\eta_4 + x\eta_6)\mathbf{j}$$
$$+ (\eta_3 + y\eta_4 - x\eta_5)\mathbf{k} \qquad (3.9)$$

We should note that η_1, η_2 and η_3 do not need to be the translatory motions of the centre of gravity of the body.

Added mass and damping terms

The added mass and damping loads are steady-state hydrodynamic forces and moments due to forced harmonic rigid body motions. There are no incident waves. However, the forced motion of the structure generates outgoing waves. The forced motion results in oscillating fluid pressures on the body surface. Integration of the fluid pressure forces over the body surface gives resulting forces and moments on the body.

By defining the force components in the x-, y- and z-direction by F_1, F_2 and F_3 and the moment components along the same axis as F_4, F_5 and F_6, we can formally write the hydrodynamic added mass and damping loads due to harmonic motion mode η_j as

$$F_k = -A_{kj} \frac{\mathrm{d}^2 \eta_j}{\mathrm{d}t^2} - B_{kj} \frac{\mathrm{d}\eta_j}{\mathrm{d}t} \qquad (3.10)$$

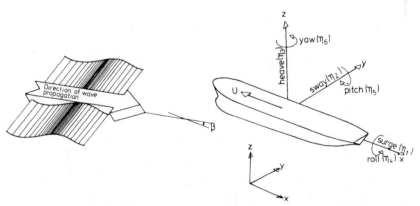

Fig. 3.2. Definitions of coordinate system, rigid-body motion modes and wave propagation direction. U is the forward speed of the ship.

A_{kj} and B_{kj} are defined as added mass and damping coefficients. There is a total of 36 added mass coefficients and 36 damping coefficients. If the structure has zero speed, half of the coefficients are zero for a structure where the submerged part has one vertical symmetry plane. For this to be true for a structure with forward speed, it is necessary that the vertical symmetry plane is parallel to the forward direction. The word added mass may be misleading, since not all of the terms have dimension of mass. Some of the terms like A_{44}, have dimensions of an inertia moment. Other terms like A_{15} have dimensions of mass multiplied by length. It can be shown that A_{kj} and B_{kj} are functions of body form, frequency of oscillation and the forward speed. Other factors like finite water depth and restricted water area will also influence the coefficients. If the structure has zero forward speed and there is no current it can be shown that $A_{kj} = A_{jk}$ and $B_{jk} = B_{kj}$.

Let us give a more detailed explanation of added mass and damping by considering forced harmonic heave motion of a structure. The heave motion causes the fluid to oscillate which means there is a pressure field in the fluid. To find the fluid motion and pressure field it is convenient to use the velocity potential. The velocity potential satisfies the Laplace equation in the fluid. In addition boundary conditions have to be imposed. On the body surface it is required that the normal component of the fluid velocity is equal to the normal component of the forced heave velocity. On the sea bed it is required that the normal component of the fluid velocity be equal to zero. Boundary conditions on the water surface are derived by requiring that the fluid pressure is equal to atmospheric pressure and that fluid particles will always remain on the free-surface (see chapter 2). At infinity a radiation condition is required. When the structure has zero forward speed and there is no current present, the far-field solution represents outgoing waves. When the velocity potential is determined, the pressure can be found by using the linearized Bernoulli's equation. Excluding the hydrostatic pressure and integrating the remaining pressure properly over the body we obtain a vertical force on the body. The linear part of this force is written as

$$F_3 = -A_{33}\frac{\mathrm{d}^2\eta_3}{\mathrm{d}t^2} - B_{33}\frac{\mathrm{d}\eta_3}{\mathrm{d}t} \tag{3.11}$$

This force is obtained by integrating the linearized pressure over the *mean* position of the body. A_{33} is added mass in heave and B_{33} is heave damping. The concept of added mass is sometimes misunderstood to be a finite amount of water which oscillates rigidly connected to the body. This is not true. The whole fluid will oscillate and with different fluid particle amplitudes throughout the fluid. In three-dimensional flow the amplitudes will always decay far away and become negligible. The added mass concept should be understood in terms of hydrodynamic pressure induced forces as above.

High-frequency limit of added mass in heave of a half circle

We will illustrate added mass in heave of circular cross-section with an axis in the free-surface by an example. Consider an infinitely long horizontal circular cylinder. In calm water the cylinder axis is at the free-surface level. We want to find the two-dimensional added mass in heave for very high frequencies ($\omega \to \infty$). By considering a very high frequency we are simplifying the problem. However, the results are useful in analysing high-frequency phenomena like ship vibrations. By two-dimensional, we mean that we study the flow in the cross-sectional plane and find forces per unit axial length of the cylinder.

In order to find added mass in heave for $\omega \to \infty$ one has to solve a boundary value problem for the velocity potential ϕ as illustrated in Fig. 3.3. From Fig. 3.3 we note that the velocity potential satisfies a two-dimensional Laplace equation in the fluid domain. On the mean wetted body surface we have the boundary condition

$$\frac{\partial \phi}{\partial r} = -\cos \theta \, |\eta_3| \, \omega \cos \omega t \quad \text{for} \quad r = R \text{ and } -\pi/2 \leqslant \theta \leqslant \pi/2$$

$$(3.12)$$

Here (r, θ) are polar coordinates and $\eta_3 = |\eta_3| \sin \omega t$ is the forced heave motion of the cylinder. Equation (3.12) approximately states that the normal component of the fluid velocity is equal to the normal component

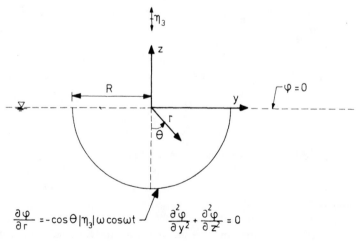

Fig. 3.3. Boundary value problem for forced heave motion $\eta_3 = |\eta_3| \sin \omega t$ of a circular cylinder at very high frequencies ω.

of the forced heave velocity on the cylinder surface. We say 'approximately' because the body boundary condition is not satisfied on the instantaneous position of the wetted body surface. However, equation (3.12) is consistent with linear theory, which implies that the heave motion is small relative to the cross-sectional dimension of the cylinder. By a Taylor expansion of the exact body boundary condition one can show that equation (3.12) is correct within linear theory.

We note that we have used the free-surface condition $\phi = 0$ on the mean free-surface (see Fig. 3.3). Since ϕ is constant on $z = 0$, this means the velocity cannot be horizontal along the free-surface and has to be vertical. The conventional linearized free-surface condition is (see equation (2.15))

$$-\omega^2 \phi + g \frac{\partial \phi}{\partial z} = 0 \quad \text{on} \quad z = 0 \tag{3.13}$$

We have neglected the second term with the gravitational acceleration g, the reason being that ω is assumed to be very high and fluid accelerations are much higher than g in the near-field of the body. If we had assumed ω to be very small, another approximation would have followed from equation (3.13), i.e. that $\partial \phi / \partial z = 0$ on $z = 0$. This is the same boundary condition we would use if there were a rigid plane at $z = 0$.

When $\omega \to 0$ or $\omega \to \infty$, the body cannot generate any free-surface waves. The reason is that the approximate free-surface conditions in these two cases state that there cannot be both a horizontal and vertical velocity component on the free-surface. Both are necessary everywhere for there to be any propagating waves.

Let us now return to the boundary value problem stated in Fig. 3.3. We will in addition assume the water depth is infinite. The velocity potential can be found by solving the 'double-body' problem in infinite fluid with no free-surface. The 'double-body' consists of the submerged body and the image body above the free-surface. The solution is

$$\phi = |\eta_3| \, \omega \cos \omega t \, \frac{R^2}{r} \cos \theta \tag{3.14}$$

We can check the solution by seeing that it satisfies the Laplace equation and necessary boundary conditions. Equation (3.14) tells us that the whole fluid oscillates harmonically. Far away from the body the oscillations are small.

The next step in finding added mass in heave is to find the pressure. Because we are considering a linear problem, the quadratic velocity term in Bernoulli's equation will be neglected. The hydrostatic pressure term $-\rho g z$ will be dealt with in the following section on restoring coefficients.

This means the dynamic pressure part of interest can be written as

$$p = -\rho \frac{\partial \phi}{\partial t} = \rho \left| \eta_3 \right| \omega^2 \sin \omega t \frac{R^2}{r} \cos \theta \tag{3.15}$$

The linear vertical force per unit length on the cylinder can be written as

$$F_3 = \int_{-\pi/2}^{\pi/2} \rho \left| \eta_3 \right| \omega^2 \sin \omega t \, R^2 \cos^2 \theta \, d\theta$$

$$= -\rho 0.5 \pi R^2 \frac{d^2 \eta_3}{dt^2} \tag{3.16}$$

According to equation (3.10) it means that the two-dimensional added mass and damping coefficients in heave are

$$A_{33} = \frac{\rho}{2} \pi R^2 \tag{3.17}$$

$$B_{33} = 0 \tag{3.18}$$

We note that the damping coefficient B_{33} is zero. This is consistent with our previous comment that the body oscillations do not generate any water waves when $\omega \rightarrow \infty$, and thus cannot carry energy away to infinity.

Energy relations

We can show by energy arguments that B_{33} is related to the wave amplitude A_3 generated by the forced heave oscillations. In order to do so we start out with general formulas presented for instance by Newman (1977: pp. 260–6). The total energy E in a fluid volume Ω consists of kinetic and potential energy. It can be written as

$$E(t) = \rho \iiint_{\Omega} (\tfrac{1}{2} V^2 + gz) \, d\tau \tag{3.19}$$

where $d\tau$ is used as a symbol for volume integration. The time derivative of the energy can be written as

$$\frac{dE(t)}{dt} = -\rho \iint_{S} \left(\frac{\partial \phi}{\partial t} \frac{\partial \phi}{\partial n} - \left(\frac{p - p_0}{\rho} + \frac{\partial \phi}{\partial t} \right) U_n \right) ds \tag{3.20}$$

where S is the boundary surface to Ω and $\partial/\partial n$ is the derivative along the normal unit vector $\mathbf{n}$ to S. (Positive direction is into the fluid domain. Note that Newman uses the opposite positive direction of $\mathbf{n}$.) U_n means the normal velocity of S and p_0 is the atmospheric pressure.

In applying equation (3.20) to our problem we will consider a two-dimensional problem and let the bounding surface consist of the wetted body surface S_B, two fixed vertical control surfaces S_∞ and $S_{-\infty}$ at respectively $y = \infty$ and $y = -\infty$, the free-surface S_F inside S_∞ and $S_{-\infty}$

and finally a surface S_0 far down in the fluid located between S_∞ and $S_{-\infty}$ (see Fig. 3.4). We can write

$$U_n = \partial\phi/\partial n \qquad \text{on } S_F \text{ and } S_B$$

$$U_n = \partial\phi/\partial n = 0 \quad \text{on } S_0$$

$$U_n = 0 \qquad\qquad \text{on } S_{\pm\infty}$$

$$p = p_0 \qquad\qquad \text{on } S_F$$

This means

$$\frac{dE}{dt} = \int_{S_B} (p - p_0) U_n \, ds - \rho \int_{S_\infty + S_{-\infty}} \frac{\partial\phi}{\partial t} \frac{\partial\phi}{\partial n} \, ds \qquad (3.21)$$

The first term on the right hand side of the equation can be written as

$$\int_{S_B} (p - p_0) U_n \, ds = \frac{d\eta_3}{dt} \int_{S_B} (p - p_0) n_3 \, ds$$

$$= \frac{d\eta_3}{dt} \left(A_{33} \frac{d^2\eta_3}{dt^2} + B_{33} \frac{d\eta_3}{dt} + C_{33}\eta_3 - \rho g V \right)$$

$$(3.22)$$

We have here used the definition of added mass and damping and included a term $C_{33}\eta_3 - \rho g V$ which arises due to the pressure term $-\rho g z$. Physically the C_{33}-term represents a change in the buoyancy of the body due to a displacement η_3 in the vertical direction. V is the

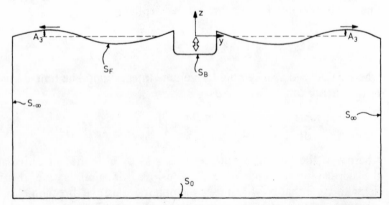

Fig. 3.4. Control surfaces used in evaluating the connection between the heave damping coefficient and the far-field wave amplitude A_3 generated by forced, harmonic and steady-state heave oscillations of a two-dimensional body. The submerged body has a vertical symmetry line.

displaced volume of the fluid when $\eta_3 = 0$. The hydrodynamic force representation is correct within linear theory.

The next step in the analysis is to average equation (3.21) over the oscillation period T. Since the energy of the fluid motion inside S is periodic, we can write

$$\int_0^T \frac{dE}{dt}\, dt = 0$$

By algebra it follows that the first term on the right hand side of equation (3.21) is

$$\int_0^T dt \left[\int_{S_B} (p - p_0) U_n\, ds \right] = \omega^2 \frac{T}{2} B_{33} |\eta_3|^2 \tag{3.23}$$

Here $|\eta_3|$ is the amplitude of the heave motion.

Physically, equation (3.23) is equal in magnitude and opposite in sign to the work done in one period by the hydrodynamic forces on the body. We note that there is no contribution from the added mass and buoyancy term in equation (3.22). In the last term of equation (3.21) we let the velocity potential represent outgoing waves. We can then write

$$\phi = \frac{gA_3}{\omega} e^{kz} \cos(\omega t \mp ky + \alpha) \tag{3.24}$$

when $y \to \pm \infty$. This assumes that the submerged body has the z-axis as a symmetry-axis. By algebra it follows that

$$\rho \int_0^T dt \int_{S_\infty} \frac{\partial \phi}{\partial t} \frac{\partial \phi}{\partial n}\, ds \simeq -\rho \int_0^T dt \int_{-\infty}^0 \frac{\partial \phi}{\partial t} \frac{\partial \phi}{\partial y}\, dz = \frac{\rho g}{2} A_3^2 \frac{g}{2\omega} T \tag{3.25}$$

The integral over $S_{-\infty}$ gives the same result as equation (3.25). This is the mean rate of energy flux through the surface S_∞ multiplied by the period T. Further, $(\rho g/2)A_3^2$ is the same as the mean energy density of the outgoing waves and $0.5g/\omega$ is the energy propagation velocity (see equation (2.19)). It now follows from time averaging of equation (3.21) that

$$B_{33} = \rho \left(\frac{A_3}{|\eta_3|} \right)^2 \frac{g^2}{\omega^3} \tag{3.26}$$

This expression is valid for any frequency. It shows that the damping coefficient can never be negative. Similar results hold for B_{ii}, $i = 2$ and 4. We do not have a similar guarantee for the added mass coefficient, which actually can be negative for certain body shapes and frequencies. Examples have been documented for catamarans, bulb sections and submerged sections close to the free-surface. Fig. 3.5 illustrates an example for sway added mass of a two-dimensional catamaran section that consists of two circular cylinders with axes in the mean free-surface and radius R. The distance between the two axes is $2p$. Large negative added mass values occur in Fig. 3.5 for frequencies close to the lowest natural frequency ω_n for antisymmetric sloshing modes between the two hulls. With sloshing we mean resonant liquid oscillations between the

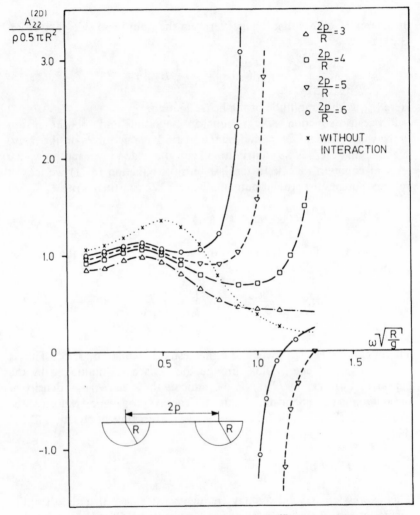

Fig. 3.5. Two-dimensional added mass in sway $A_{22}^{(2D)}$ for catamaran sections. The cross-sectional form of each hull is circular with the axis in the mean free-surface. Infinite water-depth. ($2p$ = the distance between the cylinder axes, R = Cylinder radius, ω = circular frequency of oscillation.)

two hulls. We can estimate ω_n by assuming that resonance occurs when there is half a wavelength between the inner sides of the two hulls, i.e. we can write

$$\omega_n \left(\frac{R}{g}\right)^{\frac{1}{2}} = \left(\frac{\pi}{\frac{2p}{R} - 2}\right)^{\frac{1}{2}} \tag{3.27}$$

The damping of sloshing oscillations is low since the resulting standing wave carries no energy away. This means that the wave elevations become large at the inner sides of the hulls for frequencies close to ω_n. A characteristic feature of resonant systems with low damping is that the phase angle of the response changes rapidly by 180° by varying the excitation frequency through the resonance frequency. This means that the wave amplitude at the inner side of a hull can change from being in phase with the sway acceleration to being 180° out of phase. When resonance occurs and large wave motion is present between the two hulls, the hydrodynamic pressure on the inner side of the hull is closely in phase and approximately proportional to the wave elevation in the vicinity. The hydrodynamic pressure on the outer side of the hulls is small in comparison. By remembering that added mass in sway has to do with horizontal hydrodynamic pressure forces caused by forced sway oscillation, we can understand that both very large positive and negative added mass values occur in Fig. 3.5 in the vicinity of $\omega_n (R/g)^{\frac{1}{2}}$ (see equation (3.27)).

Parameter dependence of added mass and damping
The added mass and damping coefficients may show a strong frequency dependence. As another example of this we may note that the added mass in heave for a surface-piercing two-dimensional body in deep water goes logarithmically to infinity when $\omega \to 0$.

The added mass and damping coefficients depend on the motion mode. That means that added mass in heave for a body is not necessarily the same as added mass in sway. For instance, for the circular cylinder in Fig. 3.3 it can be shown that the added mass in sway is equal to $\rho(\pi/2)R^2$ when $\omega \to 0$ while it goes logarithmically to infinity for heave when $\omega \to 0$. Fig. 3.6 illustrates the frequency dependency of the added mass and damping in heave and sway for a circular cylinder with an axis in the mean free-surface. The results are based on numerical calculations. We should note that $B_{33}^{(2D)}/(\rho \omega A)$ approaches the finite value $8/\pi$ when $\omega \to 0$. This implies that the damping coefficient goes to zero when $\omega \to 0$. As pointed out earlier this is what is expected from physical arguments.

The added moment in roll for the same cylinder depends on the choice of roll axis. If we select the roll axis to go through $(0, 0)$, there is no normal velocity component induced by the forced roll motion. This means no fluid motion, i.e. roll added moment and damping are zero in this case.

All the examples above have been two-dimensional cases. The two-dimensional added mass and damping coefficients may be combined with strip theory to obtain an approximation to the three-dimensional added mass and damping coefficients for a ship. The principle is to divide the underwater part of the ship into a number of strips (about 20). This is illustrated in Fig. 3.7. Two-dimensional coefficients are calculated for each strip and combined according to which added mass and damping coefficient is wanted. Using strip theory implies that the variation of the flow in the cross-sectional plane is much larger than the variation of the flow in the longitudinal direction. This will not be true at the ends of the body. Let us illustrate the use of strip theory to find

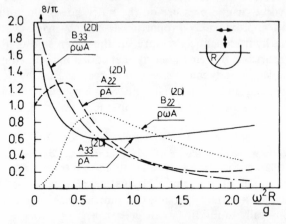

Fig. 3.6. Two-dimensional added mass and damping in heave and sway for circular cylinder with axis in the mean free-surface. Infinite water depth. ($A_{22}^{(2D)}$ = added mass in sway, $B_{22}^{(2D)}$ = damping in sway, $A_{33}^{(2D)}$ = added mass in heave, $B_{33}^{(2D)}$ = damping in heave, ρ = mass density of water, $A = 0.5\pi R^2$, ω = circular frequency of oscillation).

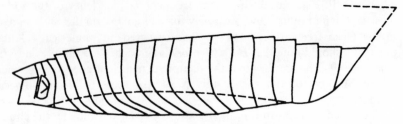

Fig. 3.7. Illustration of strip theory for ships.

added mass and damping coefficients by considering a cylinder of finite length L with constant circular cross-section of radius R. The x-axis coincides with the cylinder axis and the x-coordinates of the cylinder ends are at $\pm L/2$. The cylinder is assumed to be in infinite fluid. As an example let us consider how to find the added moment in pitch A_{55}. Following the definition of added mass we have to study forced pitch oscillations. A strip with average longitudinal coordinate x and length dx will be exposed to a vertical acceleration $-x\ddot{\eta}_5$. Associated with this acceleration there will be a vertical added mass force on the strip that is equal to $\rho\pi R^2\,dx\,x\ddot{\eta}_5$. This force creates a moment about the y-axis. By integrating over the cylinder length we find the total pitch moment to be $[-\rho\pi R^2\int_{-L/2}^{L/2}x^2\,dx]\ddot{\eta}_5$. As a matter of definition this moment is equal to $-A_{55}\ddot{\eta}_5$. This means $A_{55}=\frac{1}{12}\rho\pi R^2 L^3$. Since there is no free surface $B_{55}=0$.

The added mass and damping coefficients can be significantly influenced by the body shape. Numerical methods have to be used in a general case to estimate these coefficients (see the following chapter 4). In two-dimensional problems for ship cross-sections either source techniques or conformal mapping techniques are most commonly used. Lewis forms are examples arising from a conformal mapping technique that applies to most free-surface piercing ship cross-sections in deep water. Implicit in applying the method we say that the non-dimensionalized added mass and damping coefficients will only depend on the beam–draught ratio B/D, the sectional area coefficient $\sigma = A/(B \cdot D)$ and a non-dimensionalized frequency of oscillation like $\omega(D/g)^{\frac{1}{2}}$. Here A means the submerged cross-sectional area. This result will be used in a later section when we discuss how ship motions depend on main hull parameters.

Added mass results based on Lewis form technique are presented in Fig. 3.8. This shows the added mass in heave for infinite frequency as a function of $B/(2D)$ and $\sigma = A/(B \cdot D)$. Only the range $0.5 < \sigma < 1.0$ is shown. Outside this range both for larger and very small σ-values the Lewis form transformation represents a poor geometrical approximation of typical ship cross-sectional shapes. For cross-sections with sharp corners, for instance rectangular sections, the results are only approximate. The results in Fig. 3.8 can also be applied to heave added mass values for cross-sectional shapes in infinite fluid when the body has the y-axis as a symmetry axis. The height of the body in the oscillation direction η_3 is $2 \cdot D$ and the cross-sectional area of the body is $2 \cdot A$. The added mass-value $A_{33}^{(2D)}$ will be twice the values obtained from Fig. 3.8.

For semi-submersibles we will later calculate the response for wavelengths that are large compared to the cross-sectional dimensions of the structural parts. This implies small potential-flow damping coefficients and frequency-independent ($\omega \to 0$) added mass coefficients

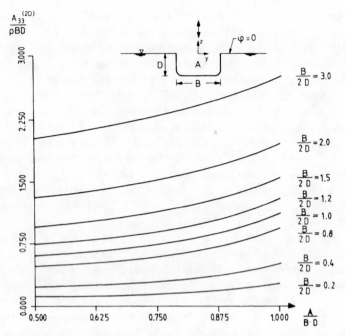

Fig. 3.8. Two-dimensional added mass in heave $A_{33}^{(2D)}$ for Lewis form sections when the frequency of oscillation $\omega \to \infty$. Infinite water depth (A = cross-sectional area, B = Beam, D = Draught, ϕ = velocity potential). The results can be applied to heave added mass for the 'double-body' with height $2 \cdot D$ and area $2 \cdot A$. The added mass will be twice the values in the figure.

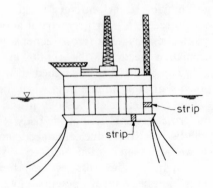

Fig. 3.9. Strip theory for semi-submersibles.

(but excluding the use of heave added mass of two-dimensional surface piercing bodies, which diverge).

A semi-submersible consists very often of long cylindrical parts. To calculate the effect of the cylindrical parts on the added mass, strip theory may be used. This is illustrated in Fig. 3.9. In strip theory it is necessary to know the two-dimensional added mass coefficients. The discussion of Fig. 3.8 should be remembered. In addition we may note that for an ellipse oscillating in sway (see Fig. 3.10), the two-dimensional added mass coefficients in sway, $A_{22}^{(2D)}$ is equal to $\rho\pi a^2$. A special case

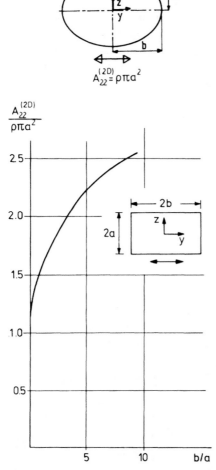

Fig. 3.10. Two-dimensional added mass coefficients in sway for an ellipse and rectangular cross-sections in infinite fluid.

of this is a circle, where the added mass is equal to the displaced mass of the fluid. Another special case is a flat plate, where we note that the added mass is infinite relative to the displaced mass. If the body is long in the oscillation direction, the added mass is much smaller than the displaced mass. We have also shown results for rectangular sections in Fig. 3.10.

The added mass value will be influenced when a cylinder comes close to the free-surface, a wall or another body. In Fig. 3.11 results are shown for a horizontal circular cylinder when $\omega \to 0$ (which is the same as long wavelengths). The plane $z = 0$ can either be understood as an idealization of the free-surface or as a wall. The distance from the top of the cylinder is called H and the added mass in heave and sway is shown as a function of H/R where R is the radius of the cylinder. We note that the closer the cylinder is to the plane $z = 0$, the greater is the added mass. When the cylinder touches the plane $z = 0$ the added mass is $(\pi^2/3 - 1)$ or 2.29 times the added mass for large values of H/R. Further, we note that the cylinder has to be quite close to $z = 0$ before there is any influence. For instance, when $H/R = 1.0$ there is only 20% increase in the added mass compared to the value for large values of H/R. Analytical solutions for this problem have been reviewed by Greenhow & Li (1987).

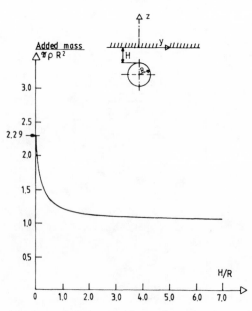

Fig. 3.11. Added mass in sway and heave for a circular section close to a wall.

Forward speed effects

A current or the forward speed of a structure will influence the added mass and damping coefficients. An important effect for a ship at forward speed is the effect of the frequency of an encounter. We will explain this by an idealistic situation that can be created in a ship model basin. In one end of the tank there is a wave maker that creates sinusoidal waves of period T_0. On the towing carriage we have mounted the vessel. The carriage is heading into the waves with a constant speed U. The phase speed of the waves (i.e. the propagation speed of the wave profile) is $c = g/\omega_0$. Let us concentrate on one point P on the vessel and consider the time T_e it takes for two successive wave crests to pass the point P. This will obviously be less than T_0. From Fig. 3.12 we can deduce

$$UT_e + cT_e = \lambda$$

where $\lambda = 2\pi g/\omega_0^2$. This means that the circular frequency of encounter $\omega_e = 2\pi/T_e$ between the ship and the waves can be written as

$$\omega_e = \omega_0 + \omega_0^2 U/g \tag{3.28}$$

For a general heading angle β between the vessel and the wave propagation direction we can write

$$\omega_e = \omega_0 + \frac{\omega_0^2 U}{g} \cos \beta \tag{3.29}$$

Here $\beta = 0°$ is head seas, $\beta = 90°$ is beam seas and $\beta = 180°$ is following seas. We note that $\omega_e = \omega_0$ for beam seas and that ω_e is smaller than ω_0 for following seas. The frequency of encounter can actually be zero which means the ship stays with the wave profile. For following seas ($\beta = 180°$) this occurs when $T_0 = 2\pi U/g$. According to equation (3.29) ω_e can be negative. If this occurs we will interpret ω_e as the absolute value of the right hand side of equation (3.29). Because the ship will

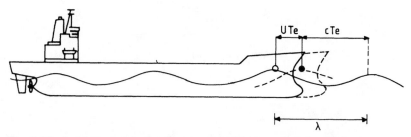

Fig. 3.12. A ship in regular head sea waves with an explanation of the frequency of encounter effect. (U = forward speed of the ship, T_e = encounter period between the ship and the waves, c = phase speed of the waves, λ = wavelength).

oscillate with the circular frequency of encounter, the added mass and damping coefficients have to be evaluated for ω_e.

In the theoretical derivation of ship motions at forward speed, the added mass and damping coefficients can be written in the form

$$A_{jk} = A_{jk}{}^0 + UA_{jk}{}^{(1)} + U^2 A_{jk}{}^{(2)}$$

$$B_{jk} = B_{jk}{}^0 + UB_{jk}{}^{(1)} + U^2 B_{jk}{}^{(2)}$$

For instance Salvesen *et al.*'s (1970) strip theory states that added mass and damping coefficients for heave and pitch of ships with pointed ends can be expressed as

$$A_{33} = \int_L A_{33}{}^{(2D)}(x)\,dx$$

$$B_{33} = \int_L B_{33}{}^{(2D)}(x)\,dx$$

$$A_{35} = -\int_L x A_{33}{}^{(2D)}(x)\,dx + \frac{U}{\omega_e{}^2} B_{33}$$

$$B_{35} = -\int_L x B_{33}{}^{(2D)}(x)\,dx - UA_{33}$$

$$A_{53} = -\int_L x A_{33}{}^{(2D)}(x)\,dx - \frac{U}{\omega_e{}^2} B_{33}$$

$$(3.30)$$

$$B_{53} = -\int_L x B_{33}{}^{(2D)}(x)\,dx + UA_{33}$$

$$A_{55} = \int_L x^2 A_{33}{}^{(2D)}(x)\,dx + \frac{U^2}{\omega_e{}^2} A_{33}$$

$$B_{55} = \int_L x^2 B_{33}{}^{(2D)}(x)\,dx + \frac{U^2}{\omega_e{}^2} B_{33}$$

One reason for the terms proportional to U and U^2, can be seen by examining the Bernoulli's equation. To show this we first write the total velocity potential as

$$\Phi = Ux + \phi$$

Here the steady velocity potential Ux represents the forward speed effect of the ship in a coordinate system moving with the forward speed of the ship. Further, ϕ is the harmonically oscillating velocity potential when the ship is forced to oscillate either in heave or pitch with the frequency of encounter.

By using Bernoulli's equation (2.4), disregarding the hydrostatic pressure term $-\rho gz$ and keeping linear terms in ϕ, it follows that the linear pressure part contributing to the added mass and damping coefficients is

$$p = -\rho \frac{\partial \phi}{\partial t} - \rho U \frac{\partial \phi}{\partial x} \qquad (3.31)$$

The pressure part proportional to U will give speed-dependent added mass and damping terms. Another reason for the speed-dependent terms come from the body boundary condition. In the body boundary condition for ϕ an additional speed-dependent term will occur when we analyse forced pitch motion. This is exemplified in Fig. 3.13. A flow has to be set up to counteract the 'incident' velocity $U\eta_5$ in the cross-sectional plane.

A more complete three-dimensional-analysis of linear wave-induced motion and loads at forward speed is complicated. For practical purposes strip theory is still recommended even if it does not properly account for all physical effects. In many cases strip theory shows good agreement with experiments. However, it is important to note its limitations. A strip theory is basically a high frequency theory. That means it is more applicable in head and bow sea waves than in following and quartering sea for a ship at forward speed. The Seakeeping Committee of the 16th ITTC reports, for instance, substantial disagreement between calculated results and experimental investigations of vertical wave loads in following waves.

It should also be noted that strip theory is a low Froude number theory. It does not properly account for the interaction between the steady wave system and the oscillatory effects of ship motions. Care should be shown in applying the theory for $Fn = U/(Lg)^{\frac{1}{2}} > \approx 0.4$. Another limitation of strip theory is the assumption of linearity between response and incident wave amplitude. This means it is questionable to apply in high sea state with ship slamming and water on deck occurring.

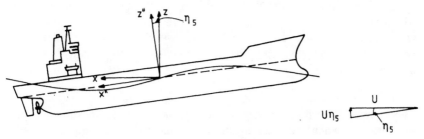

Fig. 3.13. Influence of forward speed U on the body boundary condition for a ship forced to oscillate in pitch η_5.

Strip theory is also questionable to apply for ships with low length to beam ratios. The reason is that strip theory is a slender body theory. On the other hand, the Seakeeping Committee of the 18th ITTC concludes that strip theory appears to be remarkably effective for predicting the motions of ships with length to beam ratios as low as 2.5.

Restoring forces and moments

When a body is freely floating, the restoring forces will follow from hydrostatic and mass considerations. We may write the force and moment components as

$$F_k = -C_{kj}\eta_j \tag{3.32}$$

which defines the restoring coefficients C_{kj}. The only non-zero coefficients for a body with the $x–z$ plane as a symmetry plane for the submerged volume are

$$C_{33} = \rho g A_{\mathrm{WP}}$$

$$C_{35} = C_{53} = -\rho g \iint\limits_{A_{\mathrm{WP}}} x \, ds$$

$$\tag{3.33}$$

$$C_{44} = \rho g V(z_{\mathrm{B}} - z_{\mathrm{G}}) + \rho g \iint\limits_{A_{\mathrm{WP}}} y^2 \, ds = \rho g V \overline{\mathrm{GM}}_{\mathrm{T}}$$

$$C_{55} = \rho g V(z_{\mathrm{B}} - z_{\mathrm{G}}) + \rho g \iint\limits_{A_{\mathrm{WP}}} x^2 \, ds = \rho g V \overline{\mathrm{GM}}_{\mathrm{L}}$$

Here A_{WP} is the waterplane area, V is the displaced volume of water, z_{G} and z_{B} are the z-coordinates of the centre of gravity and centre of buoyancy, respectively. $\overline{\mathrm{GM}}_{\mathrm{T}}$ is the transverse metacentric height, $\overline{\mathrm{GM}}_{\mathrm{L}}$ is the longitudinal metacentric height. We can, for instance, deduce C_{33} by considering forced heave motion and analysing the resulting changes in buoyancy forces due to the hydrostatic pressure $-\rho g z$. This can be linearly approximated as $-\rho g A_{\mathrm{w}} \eta_3$. From this C_{33} follows from equation (3.32). For a moored structure additional restoring forces have to be added. However, the effect of a spread mooring system on the linear wave-induced motion is generally quite small. In special cases, in particular for long wavelengths, the mooring system will have an influence.

Linearized wave exciting forces and moments

The wave exciting forces and moments on the structure are the loads on the structure when the structure is *restrained* from oscillating and there are incident waves. We assume the waves are regular and sinusoidal. The unsteady fluid pressure can be divided into two effects. One effect is the unsteady pressure induced by the undisturbed waves. The force due to the corresponding undisturbed pressure field is called a Froude–Kriloff force. In addition there will be a force because the structure changes this pressure field. This force is called a diffraction force and may be found in a similar way as the added mass and damping coefficients, i.e. one has to solve a boundary value problem for the velocity potential. The main difference is the boundary condition on the body, where the normal derivative of the diffraction velocity potential has to be opposite and of identical magnitude as the normal velocity of the undisturbed wave system. In this way we ensure that the normal component of the total velocity on the structure is equal to zero. Let us illustrate this by considering a vertical cylinder standing on the sea floor and penetrating the free-surface. A horizontal submerged cross-section of the cylinder is shown in Fig. 3.14. The figure defines polar coordinates such that $x = r \cos \theta$, $y = r \sin \theta$. The unit normal vector $\mathbf{n}$ is written as $\mathbf{n} = \cos \theta \mathbf{i} + \sin \theta \mathbf{j}$. The undisturbed pressure, i.e. the Froude–Kriloff pres-

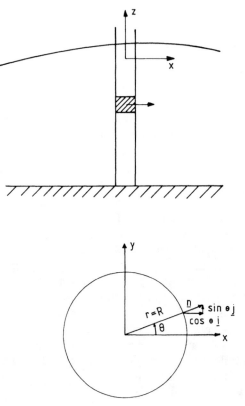

Fig. 3.14. Horizontal submerged cross-section of a vertical cylinder.

sure, can be written as (see Table 2.1):

$$p = \rho g \zeta_a \frac{\cosh k(z + h)}{\cosh kh} \sin(\omega t - kx)$$

for waves propagating along the positive x-axis. The Froude–Kriloff force on the element dz can be written as

$$dF = -i\rho g \zeta_a \frac{\cosh k(z + h)}{\cosh kh} dz\, R$$

$$\times \int_0^{2\pi} \sin(\omega t - kR \cos \theta) \cos \theta\, d\theta$$

We will now assume that the wavelength is much larger than the radius R of the cylinder. This implies that kR is small. We can write

$$dF \approx -i\rho g \zeta_a \frac{\cosh k(z + h)}{\cosh kh} dz\, R \left\{ \int_0^{2\pi} \cos \theta\, d\theta \sin \omega t \right.$$

$$\left. - \int_0^{2\pi} kR \cos^2\theta\, d\theta \cos \omega t \right\}$$

We may note that the large part of the pressure, i.e. the part that is in phase with $\sin \omega t$, does not give any contribution to the horizontal force.

We can now write

$$dF = i\rho g \zeta_a k\pi R^2\, dz \frac{\cosh k(z + h)}{\cosh kh} \cos \omega t = i\rho \pi R^2\, dz\, a_1|_{x=0}$$

Here $\rho \pi R^2\, dz$ is the mass of the displaced fluid of the cylinder strip and $a_1|_{x=0}$ is the x-component of the fluid acceleration at $x = 0$ at the average z-coordinate for the strip as *if the cylinder were not there.*

The undisturbed pressure force is only one part of the total force on the strip. We can understand this by studying the undisturbed velocity field. This causes a fluid transport through the cylinder wall. This is unphysical and the cylinder must therefore set up a pressure field which causes a velocity field that counteracts the normal component of the undisturbed velocity field at the cylinder wall. To find the diffraction force due to the additional pressure distribution we may argue as follows. Due to the long wavelength assumption we may say that the problem of finding the additional pressure distribution is equivalent to a problem where the cylinder is forced to oscillate with a velocity $-u|_{x=0}$, where $u|_{x=0}$ is the x-component of the fluid velocity at $x = 0$ at the average z-coordinate for the strip as *if the cylinder were not there.* In this way there will be no resultant fluid transport through the cylinder wall. From basic hydrodynamics we know that the force on the body due to the forced oscillatory velocity $-u|_{x=0}$ can be written as

$$A_{11} a_1|_{x=0} \mathbf{i}$$

where A_{11} is the proper added mass, which for the circular cylinder is equal to the mass of the displaced fluid volume.

What we have shown above is a derivation of the mass force in Morison's equation based on potential theory. With mass force we mean a force that is proportional to the undisturbed fluid acceleration a_1. The horizontal force per unit length on a strip of the cylinder can, according to Morison's equation (Morison et al., 1950) be written as

$$dF = \rho \pi \frac{D^2}{4} C_M a_1 + \frac{\rho}{2} C_D D \, |u| \, u \qquad (3.34)$$

Positive force direction is in the wave propagation direction. Further, ρ is the mass density of the water, D is the cylinder diameter, u and a_1 are horizontal undisturbed fluid velocity and acceleration at the mid-point of the strip. In reality the mass and drag coefficients C_M and C_D have to be empirically determined and are dependent on many parameters like Reynolds number, Keulegan–Carpenter number, a relative current number and surface roughness ratio. This will be discussed in detail in chapter 7. In this chapter the non-linear drag term is assumed to be negligible. What we have shown is that potential theory gives $C_M = 2$ for a circular cylinder, where half the contribution comes from the Froude–Kriloff force and the other half comes from the diffraction force. If viscous effects are accounted for, C_M will differ from 2. We should remember that the wavelength was assumed to be large relative to the diameter. For arbitrary wavelength we may use the linear analytical solution by McCamy & Fuchs (1954). The theory is based on potential flow and is valid for a circular cylinder standing on the sea floor and penetrating the free-surface. This theory shows that there is one force part in phase with a_1 and another force part in phase with the velocity u.

We may generalize the derivation above for any small-volume structure. By small volume we mean that the wavelength λ is large relative to a characteristic cross-sectional dimension of the body. For instance, for the vertical cylinder it means that $\lambda > 5D$ where D is the cylinder diameter. The force on a relatively small body can be written as

$$\mathbf{F} = \mathbf{i}F_1 + \mathbf{j}F_2 + \mathbf{k}F_3 \qquad (3.35)$$

where

$$F_i = -\iint_S p n_i \, ds + A_{i1} a_1 + A_{i2} a_2 + A_{i3} a_3 \qquad (3.36)$$

Here p is the pressure in the undisturbed wave field and $\mathbf{n} = (n_1, n_2, n_3)$ is the unit vector normal to the body surface defined to be positive into the fluid. The integration is over the average wetted surface of the body. Further, a_1, a_2 and a_3 are acceleration components along the x-, y- and

z-axes of the undisturbed wave field and are to be evaluated at the geometrical mass centre of the body.

The first term in equation (3.36) is the Froude–Kriloff force. The other terms physically represent the fact that the undisturbed pressure-field is changed due to the presence of the body (diffraction force).

If the body is totally submerged, the body is small relative to the wavelength and the *whole* body surface is wetted we can show that:

$$-\iint\limits_{S} p \begin{bmatrix} n_1 \\ n_2 \\ n_3 \end{bmatrix} \mathrm{d}s = \rho V \begin{bmatrix} a_1 \\ a_2 \\ a_3 \end{bmatrix} \tag{3.37}$$

where V is the displaced volume. Equation (3.37) is according to the assumptions not valid for a structure (for instance a tank) standing on the sea bed where the bottom of the structure is not wetted. However, in the case of a horizontal sea bed, the fact there is no fluid pressure on the bottom of the structure does not affect the horizontal forces. Equation (3.37) is therefore only invalid for the vertical force in this case.

Equation (3.36) is often used together with strip theory to predict wave excitation forces on semi-submersibles and TLPs when the wavelength is large relative to the cross-sectional dimensions. Special care should be taken at junctions between columns and pontoons in evaluating the Froude–Kriloff force. Equation (3.37) cannot be used in this case. We then have to integrate the Froude–Kriloff pressure force directly.

We will illustrate this by considering the vertical excitation force $F_3(t)$ on a TLP (see Fig. 3.15) for long wavelengths. A similar analysis applies for a conventional semi-submersible drilling platform. A main contribution to $F_3(t)$ comes from the pressure forces acting on the pontoons. A strip theory approach may be used to estimate these pressure forces because the pontoons can be treated as slender members. Then, the vertical wave excitation load on a strip of length, $\mathrm{d}s$, that is remote from the columns can be written as:

$$F_3 = (\rho A_\mathrm{p} + A_{33}^{(2\mathrm{D})})\,\mathrm{d}s\, a_3 \tag{3.38}$$

where A_p is the cross-sectional area, $A_{33}^{(2\mathrm{D})}$ is the two-dimensional added mass in heave in an infinite fluid, and a_3 is the vertical undisturbed fluid acceleration at the geometrical centre of the cross-sectional area. Equation (3.38) is based on a long wavelength assumption relative to the cross-section. The first term represents the Froude–Kriloff force (i.e. the pressure force on the cross-section due to the undisturbed fluid flow). It should be stressed that in equation (3.38) the Froude–Kriloff representation is valid only when the whole circumference of the cross-section is wetted. When this is not satisfied, the contribution to the Froude–Kriloff force should be obtained by properly integrating the undisturbed pressure p.

To derive an expression for $F_3(t)$, we will write

$$a_3 = -\omega^2 \zeta_a e^{kz} \sin(\omega t - kx) \tag{3.39}$$

$$p = \rho g \zeta_a e^{kz} \sin(\omega t - kx) \tag{3.40}$$

where ζ_a is the incident wave amplitude. On the pontoons oriented in the x-direction we obtain the following vertical force contribution.

$$-\omega^2 \zeta_a e^{kz_m} 2(\rho A_p + A_{33}^{(2D)}) \int_{-B/2}^{B/2} \sin(\omega t - kx)\,dx$$

where

$$\int_{-B/2}^{B/2} \sin(\omega t - kx)\,dx = (2/k) \sin \omega t \sin kB/2$$

and z_m is the z-coordinate of the geometrical centre of the cross-sectional area. B is explained in Fig. 3.15. On the pontoons oriented in the y-direction we obtain the following vertical force contribution.

$$-\omega^2 \zeta_a e^{kz_m}(\rho A_p + A_{33}^{(2D)})B(\sin(\omega t - kL_1/2) + \sin(\omega t + kL_1/2))$$

where L_1 is explained in Fig. 3.15. By using only the Froude–Kriloff

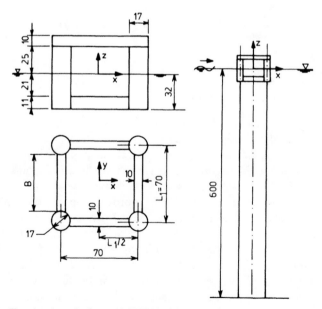

Fig. 3.15. Tension leg platform (ASM600). Dimensions are given in metres. Its main characteristics are presented in Table 5.1.

force on the columns it now follows that

$$F_3(t) = \rho g \zeta_a e^{k z_n} \sin \omega t \left[-\left(4 \sin \frac{kB}{2} + 2kB \cos \frac{kL_1}{2}\right) \right.$$

$$\left. \times \left(A_p + \frac{A_{33}^{(2D)}}{\rho}\right) + A_w e^{k(z_B - z_m)} \cos \frac{kL_1}{2} \right] \qquad (3.41)$$

where z_B is the z-coordinate of the bottom of the platform, and A_w is the total waterplane area of the platform. $A_{33}^{(2D)}$ can, for instance, be calculated by the Frank close fit method (see Frank, 1967). Strictly speaking, we should also introduce a correction to the added mass due to the columns but this is not possible to do in a simple manner and will be neglected herein.

We note that the force is either in-phase or 180° out-of-phase with the undisturbed wave elevation at $x = 0$. Furthermore, the excitation force can be zero for certain frequencies. For long wavelengths, this happens when the last term in the brackets cancels the first two terms in the brackets. Physically it means that the force on the columns cancels the force on the pontoons.

The most convenient way to calculate the wave excitation loads on a ship is to use a coordinate system fixed with respect to the forward speed of the ship. We can then write the incident wave potential as

$$\phi_I = \frac{g \zeta_a}{\omega_0} e^{kz} \cos(\omega_e t - kx \cos \beta - ky \sin \beta) \qquad (3.42)$$

where $k = \omega_0^2/g$. By using the linearized Bernoulli's equation (similar to equation (3.31)) we find that the dynamic pressure is

$$p_I = \rho g \zeta_a e^{kz} \sin(\omega_e t - kx \cos \beta - ky \sin \beta) \qquad (3.43)$$

By the nature of the Froude–Kriloff force its amplitude has to be independent of the ship's speed. This is consistent with equation (3.43). However, the amplitude of the diffraction force will in general be speed-dependent. For more details as to how the diffraction force can be calculated within a strip theory approximation one may, for instance, consult Salvesen et al. (1970). Their approach is based on using Green's second identity. The formulas use the velocity potentials due to forced motion of the structure instead of the diffraction potential of the structure. The advantage of introducing the velocity potentials due to forced motion is that it is generally more correct to use a strip theory approximation and neglect hydrodynamic interaction between the cross-sections for the forced-motion potentials than for the diffraction potentials. We illustrate this by considering head sea incident waves on a ship. As the waves propagate along the ship they will be continuously modified. This is particularly true if the wavelength is small relative to the ship length (Faltinsen, 1972). This means hydrodynamic interaction between the cross-sections has to be accounted for when the diffraction potential is solved. We should note that the use of Green's second identity to calculate wave excitation forces and moments implies that we do not find the pressure distribution along the ship.

If we want to apply an approximate method valid for long wavelengths relative to the cross-section of the ship, we could argue as we did for the vertical cylinder in the beginning of this section. We will show this for zero ship speed. The problem of finding the vertical diffraction force on a cross-section (strip) of a ship is equivalent to a problem where the strip is forced to oscillate with a vertical velocity

$$-w = -\omega \zeta_a e^{k z_m} \cos(\omega t - kx \cos \beta)$$

This velocity is minus the incident vertical velocity at the area centre $(x, 0, z_m)$ of the strip. From the section on added mass and damping we know that the vertical hydrodynamic force on the strip due to the forced velocity $-w$ is

$$A_{33}^{(2D)} a_3 + B_{33}^{(2D)} w$$

where

$$-a_3 = \omega^2 \zeta_a e^{k z_m} \sin(\omega t - kx \cos \beta)$$

This is a force per unit length. Since the body is restrained from oscillating, we should not include any restoring forces due to buoyancy. The added mass and damping coefficients are frequency dependent. Due to the long wavelength assumption, the damping term is small compared to the added mass term. However, we must not evaluate $A_{33}^{(2D)}$ for zero frequency. The reason is that $A_{33}^{(2D)}$ diverges for a surface piercing body when the frequency goes to zero.

By properly integrating the diffraction loads over the length L of the ship and including the Froude–Kriloff force we can write the vertical excitation force F_3 and the pitch moment F_5 as

$$\begin{bmatrix} F_3(t) \\ F_5(t) \end{bmatrix} = -\iint_{S_B} p_I \begin{bmatrix} n_3 \\ -xn_3 \end{bmatrix} ds$$
$$-\int_L dx \begin{bmatrix} 1 \\ -x \end{bmatrix} \omega \zeta_a e^{k z_m} \{A_{33}^{(2D)} \omega \sin(\omega t - kx \cos \beta)$$
$$-B_{33}^{(2D)} \cos(\omega t - kx \cos \beta)\} \tag{3.44}$$

Here p_I is given by equation (3.43) and S_B is the mean wetted body surface.

Newman (1962) has derived simple formulas for beam sea incident waves on an infinitely long horizontal cylinder and for incident waves on a vertical axisymmetric body. The formulas are valid for any frequency, but assumes the structure has zero forward speed and no current. The derivation is based on Green's second identity. In that way the velocity potential for forced motion of the body is introduced mathematically. Newman is able to relate the force component number i to the damping coefficient B_{ii}. For an infinitely long cylinder in beam sea, Newman (1962) writes the exciting force amplitude $|F_i|$ per unit length as

$$|F_i| = \zeta_a \left(\frac{\rho g^2}{\omega} B_{ii}^{(2D)} \right)^{\frac{1}{2}}, \qquad i = 2, 3, 4 \tag{3.45}$$

where $B_{ii}^{(2D)}$ is the two-dimensional damping coefficient in mode i, $|F_2|$ is the horizontal force amplitude, $|F_3|$ is the vertical force amplitude and $|F_4|$ is the roll moment amplitude.

For an axisymmetric body we can write

$$|F_i| = \zeta_a \left(\frac{\rho g^3}{\omega^3} 2AB_{ii} \right)^{\frac{1}{2}}, \qquad i = 1, \ldots, 5 \tag{3.46}$$

where $A = 1$ for $i = 3$ (heave) and equal to 2 for $i = 1, 2, 4, 5$. Newman's formula does not tell us what the phase of the wave excitation loads are relative to the incident waves.

For a general body shape, numerical techniques have to be used to evaluate wave excitation loads (see chapter 4). This is based on properly integrating Froude–Kriloff and diffraction pressure forces. If one only wants to find the wave excitation loads and not the detailed pressure distribution, the Haskind relation can be used (Newman, 1977). In its original form it is assumed that the structure has zero forward speed. There is no current present. The derivation is based on Green's second identity. The formulas use the velocity potentials due to forced motion of the structure instead of the diffraction potential of the structure. Equations (3.45) and (3.46) can be derived from the Haskind relations. An advantage of the Haskind relations is that they provide an independent test that the wave excitation loads are correctly calculated by a computer program.

The equations of motions
When the hydrodynamic forces have been found it is straightforward to set up the equations of rigid body motions. This follows by using the equations of linear and angular momentum. For steady-state sinusoidal motions we may write

$$\sum_{k=1}^{6} [(M_{jk} + A_{jk})\ddot{\eta}_k + B_{jk}\dot{\eta}_k + C_{jk}\eta_k] = F_j e^{-i\omega_e t} \quad (j = 1, \ldots, 6) \tag{3.47}$$

where M_{jk} are the components of the generalized mass matrix for the structure and F_j are the complex amplitudes of the exciting forces and moment-components with the force and moment-components given by the real part of $F_j e^{-i\omega_e t}$ (i is complex unit). The equations for $j = 1, 2, 3$ follows from Newton's law. For instance, let us consider $j = 1$. For a structure that has lateral symmetry (symmetric about the x–z plane) and with the centre of gravity at $(0, 0, z_G)$ in its mean oscillatory position, we can write the linearized acceleration of the centre of gravity in the x-direction as

$$\frac{d^2\eta_1}{dt^2} + z_G \frac{d^2\eta_5}{dt^2}$$

From this the components of the mass matrix M_{jk} follow as

$$M_{11} = M, \qquad M_{12} = 0, \qquad M_{13} = 0, \qquad M_{14} = 0,$$

$$M_{15} = Mz_G, \qquad M_{16} = 0$$

Here M is the mass of the structure. Similarly for $j = 2, 3$. For $j = 4, 5, 6$ we have to use the equations of angular momentum. We can then set up the following mass matrix

$$M_{jk} = \begin{bmatrix} M & 0 & 0 & 0 & Mz_G & 0 \\ 0 & M & 0 & -Mz_G & 0 & 0 \\ 0 & 0 & M & 0 & 0 & 0 \\ 0 & -Mz_G & 0 & I_4 & 0 & -I_{46} \\ Mz_G & 0 & 0 & 0 & I_5 & 0 \\ 0 & 0 & 0 & -I_{46} & 0 & I_6 \end{bmatrix} \tag{3.48}$$

where I_j is the moment of inertia in the jth mode and I_{jk} is the product of inertia with respect to the coordinate system (x, y, z).

For a structure with no forward speed and where there is no current present it can be shown by Green's second identity that the added mass and damping coefficients satisfy the symmetry relations $A_{jk} = A_{kj}$, $B_{jk} = B_{kj}$.

For a structure with lateral symmetry, the six coupled equations of motions reduce to two sets of equations, one set of three coupled equations for surge, heave and pitch and another set of three coupled equations for sway, roll and yaw. Thus for a structure with lateral symmetry, surge, heave and pitch are not coupled with sway, roll and yaw. This will be assumed in the following text. In order to understand why surge, heave and pitch are not coupled with sway, roll and yaw, we need to study the added mass and damping coefficients. Let us illustrate this by an example and analyse coupled added mass A_{23} between sway and heave for the case presented in Fig. 3.3, i.e. we consider forced heave motion of a circular cylinder when $\omega \to \infty$. The pressure on the body is given by equation (3.15). This pressure is symmetric about the z-axis. A consequence of this is that the resulting horizontal force is zero. By definition (see equation (3.10)) the added mass and damping coefficients A_{23} and B_{23} are zero. For forced heave motion of a general body shape where the submerged body surface is symmetric about the x–z plane we will also find that the pressure distribution is symmetric about the x–z plane. This means no horizontal force and no roll and yaw moment. A consequence from equation (3.10) is that A_{k3} and B_{k3} for $k = 2, 4$ and 6 are zero. A similar argument can be shown for forced pitch and surge motion, i.e. that A_{k1}, A_{k5}, B_{k1} and B_{k5} are zero for $k = 2$, 4 and 6. If we consider forced sway, roll and yaw motion, the pressure distribution will be antisymmetric about the x–z plane. This implies A_{1k}, A_{3k}, A_{5k}, B_{1k}, B_{3k} and B_{5k} are equal to zero for $k = 2$, 4 and 6.

We can combine this information with the fact that there are no coupling effects between the lateral motions (sway, yaw, roll) and the vertical and longitudinal motions (heave, pitch, surge) due to restoring or mass terms (see equations (3.33) and (3.48)). We can then conclude from the equations of motion (equation (3.47)) that the lateral motions are uncoupled from the vertical and longitudinal motions.

The equations of motions (3.47) can be solved by substituting $\eta_k = \bar{\eta}_k \, e^{-i\omega_e t}$ in the left hand side. Here $\bar{\eta}_k$ are the complex amplitudes of the motion modes. Dividing by the factor $e^{-i\omega_e t}$ the resulting equations can be separated into real and imaginary parts. This leads to six coupled algebraic equations for the real and imaginary parts of the complex amplitudes for surge, heave and pitch. A similar algebraic equation system can be set up for sway, roll and yaw. These matrix equations can be solved by standard methods. When the motions are found, the wave loads can be obtained using the expressions for hydrodynamic forces which we discussed previously.

It should be stressed that equations (3.47) are only generally valid for steady state sinusoidal motions. For instance, in a transient free-surface problem the hydrodynamic forces include memory effects and do not depend only on the instantaneous values of body velocity and acceleration (Ogilvie, 1964).

DISCUSSION ON NATURAL PERIODS, DAMPING AND EXCITATION LEVEL

The natural or resonance periods, damping level and wave excitation level are important parameters in assessing the amplitudes of motion of a platform or a vessel. Relatively large motions are likely to occur if the structures are excited with oscillation periods in the vicinity of a resonance period. However, if the damping is high or the excitation level is relatively low due to cancellation effects, it may be difficult to distinguish the response at resonance periods from the response at other periods.

The uncoupled and undamped resonance periods can be written as

$$T_{ni} = 2\pi \left(\frac{M_{ii} + A_{ii}}{C_{ii}} \right)^{\frac{1}{2}}$$ (3.49)

For an unmoored structure there are no (uncoupled) resonance periods in surge, sway and yaw. For a typical moored structure the natural periods in surge, sway and yaw are of the order of magnitude of minutes and will therefore be long relative to wave periods occurring in the sea. We will see in chapter 5 that non-linear effects may excite resonant oscillations at these long periods.

The natural period in heave for a semi-submersible or a ship, or any other type of freely floating body can be written as

$$T_{n3} = 2\pi \left(\frac{M + A_{33}}{\rho g A_{w}} \right)^{\frac{1}{2}}$$ (3.50)

where A_w is the waterplane area. It is common design procedure for semi-submersibles to require that the natural periods in heave, pitch and roll are larger than $T = 20\,s$, i.e. high relative to most wave periods occurring in the open sea. This is possible to achieve by the low water-plane area of semi-submersibles. For a ship we can write equation (3.50) as

$$T_{n3} = 2\pi \left(\frac{C_B}{C_w} \frac{D}{g} \left(1 + \frac{A_{33}}{M} \right) \right)^{\frac{1}{2}}$$

where $C_B = V/(BLD)$ ($V =$ displaced volume of the ship) is the block coefficient and $C_w = A_w/(BL)$ the waterplane area coefficient. If the ship does not have forward speed, the heave oscillations at the natural period will be excited by waves with wavelength

$$\lambda = \frac{g}{2\pi} T_{n3}^2 = 2\pi \frac{C_B}{C_w} D \left(1 + \frac{A_{33}}{M} \right)$$ (3.51)

Since $2\pi D$ is in the range of the ship length, equation (3.51) means that the resonant wavelength is of the same order of magnitude as the ship length. If the ship has a forward speed it is a different wavelength that causes the resonant heave oscillations. The requirement for resonance is that the encounter period T_e is equal to T_{n3}. This means

$$\frac{T_0}{1 + \dfrac{2\pi}{T_0} \dfrac{U}{g} \cos \beta} = T_{n3}$$ (3.52)

This is illustrated in Fig. 3.16. The figure tells for instance that period

$T_0 = 12$ s causes heave resonance when the vessel is heading with a speed 10 m s^{-1} (19.4 knots) against the waves. At zero speed, resonance is caused by a period of 7.8 s. For head seas it means that the wavelength that creates resonance increases with the forward speed. If the vessel changes direction, for instance to bow sea ($\beta = 45°$ heading), a resonant heave oscillation will not occur for a speed of 10 m s^{-1} and wave period $T_0 = 12$ s. It would occur if the ship had a speed of 10 m s^{-1}/cos $45° = 14$ m s^{-1}.

The natural period in heave for a tension leg platform can be written as

$$T_{n3} = 2\pi \left(\frac{M + A_{33}}{EA/l} \right)^{\frac{1}{2}} \tag{3.53}$$

where E, A and l are respectively modulus of elasticity, cross-section area of the tendons and length of the tendons. The stiffness due to the waterplane area is negligible in comparison with the restoring effect of the tendons. We note from equation (3.53) that the heave natural period increases with increasing tendon length (i.e. the water depth). Similar considerations apply for the pitch and roll natural periods. Generally speaking, the natural periods in heave, pitch, and roll of a TLP will be low relative to most wave periods occurring in the open sea. However, they may be excited by non-linear second-order effects (see chapter 5).

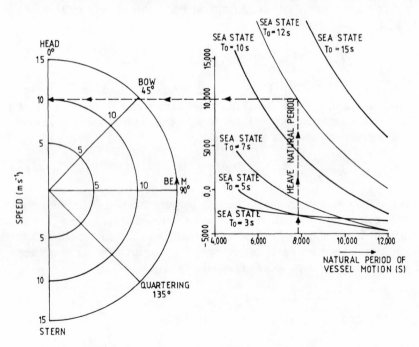

Fig. 3.16. Relation between resonance period, wave period, forward speed and wave propagation direction.

The uncoupled natural period in pitch for a freely floating body like a ship or a semi-submersible can be written as

$$T_{n5} = 2\pi \left(\frac{Mr_{55}^2 + A_{55}}{\rho g V \overline{GM}_L} \right)^{\frac{1}{2}} \tag{3.54}$$

where r_{55} is the pitch radius of gyration with respect to an axis parallel with the y-axis through the centre of gravity, A_{55} is the pitch added moment and $\overline{GM}_L$ is the longitudinal metacentric height. For a ship, r_{55} can be approximated as 0.25 times the ship length. The order of magnitude of T_{n5} for a ship is the same as T_{n3}.

The uncoupled natural period in roll is

$$T_{n4} = 2\pi \left(\frac{Mr_{44}^2 + A_{44}}{\rho g V \overline{GM}_T} \right)^{\frac{1}{2}} \tag{3.55}$$

where r_{44} is the roll radius of gyration with respect to an axis parallel with the x-axis through the centre of gravity, A_{44} is the roll added moment and $\overline{GM}_T$ is the transverse metacentric height. For a ship, r_{44} is typically 0.35 times the beam. The transverse metacentric height depends on the loading condition. A design rule/recommendation for a ship is to try to select natural periods in roll larger than 10 s. In this way rolling is not a problem in small and moderate sea states. The parameter which has most influence on the natural period is the metacentric height. T_{n4} is typically 4–6 s for small fishing vessels, 8–12 s for conventional merchant vessels and up to 20–25 s for specialized heavy lift vessels. Semi-submersibles have T_{n4} in the range of 35–50 s. T_{n4} depends very much on the stability requirements and how the ship is built with respect to damage stability.

Ships without roll stabilization equipment are exposed to strong resonance effects in roll. The amplitude at resonance depends on the damping level. The damping in roll for a ship at zero Froude number, i.e. zero forward speed, is due to wave generation, viscous effects and roll stabilization equipment. At high Froude numbers the lifting effect of the hull and the rudders is important. By lift we do not mean that there is a vertical force, but that the ship and the rudders act as lifting surfaces in a hydrodynamic sense. The roll wave damping is due to the waves created by the roll motion and is frequency dependent. When the frequency goes to zero or infinity, the roll damping goes to zero. Since the roll damping dominates the roll motion around roll resonance, the natural period in roll will have an important influence on the effect of the roll wave damping. The roll wave damping has a tendency to be small due to cancellation effects for normal midship sections. By cancellation we mean that the roll moment caused by the pressure forces on the ship sides tend to counteract the roll moment caused by the pressure forces on the ship bottom. For a circular cross-section the roll damping will be zero if the moment axis coincides with the cylinder axis. If either the cross-sectional beam is small or large, the cancellation effect will be less pronounced. This will cause relatively large roll wave damping. This is illustrated in Fig. 3.17 where two-dimensional roll wave damping coefficients $B_{44}^{(2D)}$ for rectangular cross-sections are shown for different oscillation periods T as a function of the beam draught ratio B/D. The oscillation periods in Fig. 3.17 are in a limited range. This means the data do not show that $B_{44}^{(2D)} \to 0$ when $T \to 0$ or ∞.

Viscous effects can be divided into skin friction effects and viscous effects due to the pressure distribution around the ship. The last effect is often associated with eddy making and is therefore called eddy-making damping in the literature.

The effect of skin friction is more important in model scale than in full scale. In model scale it cannot be completely neglected, while it is negligible in full scale. Depending on the frequency and the roll amplitude, the roll wave damping may be larger or smaller than the eddy-making damping (Ikeda *et al.*, 1977a). For a ship without bilge

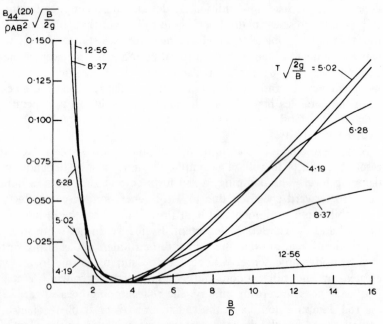

Fig. 3.17. Two-dimensional roll wave damping $B_{44}^{(2D)}$ as a function of beam–draught-ratio for rectangular cross-sections. Infinite water depth. The roll axis is in the mean free-surface. (B = beam, D = Draught, A = submerged cross-sectional area, T = period of oscillations.) (Vugts, 1968.)

keels, the bilge radius has an important influence on the eddy-making damping for midship sections (Tanaka, 1961). The eddy-making damping can be quite large for rectangular cross-sections while it is of less importance for conventional midship sections. Kato (1966) and Ikeda *et al.* (1977b) have given empirical formulas for roll damping due to bilge keels. The bilge keel damping can very well amount to 50% of the total damping. Eddy-making and bilge keel damping is non-linear and expressed in the form of $B\dot{\eta}_4|\dot{\eta}_4|$, where dots stand for time derivatives. In practical calculations equivalent linearization is used to determine the roll amplitudes (see exercise 3.5).

Heave and pitch damping of monohull ships are dominated by wave damping. For catamarans at zero or very low forward speeds small wave damping may occur due to interaction effects of the two hulls. This occurs typically when the distance between the hulls is an odd number of half-wavelengths. A simple way of explaining the phenomenon is to combine the far-field wave picture generated by each hull and neglect the diffraction caused by the other hull. We will illustrate what we mean by considering a two-dimensional case. The wave elevation in the far-field ($y \rightarrow +\infty$) due to forced heave motion of the two hulls is written as

$$A_3 \sin(\omega t - k(y - p) + \alpha) + A_3 \sin(\omega t - k(y + p) + \alpha)$$
$$= 2A_3 \cos(kp) \sin(\omega t - ky + \alpha) \quad (3.56)$$

Here A_3 and α are the wave amplitudes and phases caused by one hull as if the other were not there. Further, $2p$ is the distance between the centre planes of each hull. We note that equation (3.56) becomes equal to zero when $\cos(kp) = 0$, i.e. $\lambda/(2p) = 2/(2n + 1)$, $n = 0, 1, \ldots$. The most interesting practical case is when $n = 0$, i.e. when there is half a wavelength between the centre planes of the two hulls. When the far-field wave amplitude generated by forced heave oscillation is zero we know from equation (3.26) that the damping in heave is zero. Both three-dimensional flow effects and forward speed effects will cause less cancellation of the far-field wave system around a catamaran. This is particularly true for a high-speed catamaran.

Small excitation forces may also occur due to cancellation effects. One example of this is a ship in head sea waves with a wavelength of the order of the ship length. The phase difference in the vertical excitation loads along the ship causes the total heave excitation force to be small. Since the heave resonance period for a ship at zero forward speed corresponds to a wavelength of the order of the ship length, the heave motion may be rather small at the heave resonance.

For semi-submersibles and TLPs, the Froude–Kriloff force and the diffraction force tend to cancel each other at large periods, typically between 15 and 20 s. This is evident from Fig. 3.18.

LINEAR WAVE-INDUCED MOTIONS AND LOADS ON A TENSION LEG PLATFORM (TLP) IN THE MASS-FORCE DOMAIN

Fig. 3.18 presents results of calculated linear motions for the ASM600 platform (Fig. 3.15) in head seas. These calculations are based on incident regular waves using the theory described by Faltinsen *et al.* (1982). Here we will only try to explain the essential part of the calculations when wave diffraction is of small importance, and focus our attention on the heave motion η_3 of the centre of gravity of the platform.

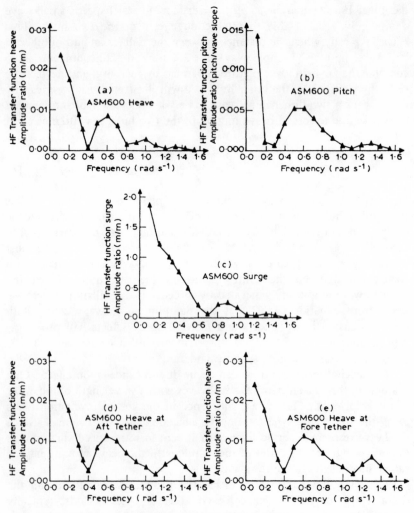

Fig. 3.18. Motion of the TLP presented in Fig. 3.15.

The undamped equation of motion can be written as

$$(M + A_{33}) \frac{d^2 \eta_3}{dt^2} + \frac{EA}{l} \eta_3 = F_3(t) \qquad (3.57)$$

where the dominating restoring coefficient C_{33} is due to tethers. E is the modulus of elasticity, A is the total cross-sectional area and l is the length of the tendons.

The frequencies of interest are much lower than the resonance frequency in heave. We will therefore neglect the first term on the left hand side. A formula for the excitation force $F_3(t)$ has been derived in equation (3.41). It was noted that the force is either in-phase or 180° out-of-phase with the undisturbed wave elevation at $x = 0$. Furthermore, the excitation force can be zero for certain frequencies. This is also evident from Fig. 3.18(a). The force is not exactly zero in the figure because wave diffraction was also incorporated in the computer program. The small vertical force at the largest period in Fig. 3.18(a) occurs because the forces on the columns cancel the force contribution from the pontoons.

Computer calculations have shown that the volumetric ratio, R_v, of columns to pontoons is an important parameter for the dynamic tension variation. We can write

$$R_v \approx \frac{A_w}{A_p} \frac{D}{4B} \qquad (3.58)$$

where D is the draught of the platform. This can partly be seen from equations (3.41) and (3.57). The ratio A_w/A_P is a parameter which controls the cancellation period of (3.41), and D governs how much wave energy there is at the pontoon depth. The spacing between the pontoons, which is related to B, will also influence the dynamic tensions according to equation (3.57).

The calculation of the pitch and surge responses in Fig. 3.18 can be based on a similar procedure to that outlined for heave. In this case the columns will also contribute to the wave excitation force. The heave at aft or fore tendons can be calculated by properly combining the heave at the center of gravity and the pitch motion. Fig. 3.18 indicates that the 'design wave approach' may be dangerous to apply in predicting the tendon loads in a survival situation. These loads are sensitive to the frequency range of interest. By 'design wave approach' we mean that we only consider the response due to one regular wave system. Let us say we chose a wave period in the vicinity of 15 s ($\omega = 0.42$ rad/s). We see from Fig. 3.18 that the tendon loads are sensitive to what wave period we select.

Standard procedures exist for calculating irregular sea effects when the response is linear (see the introduction to this chapter). The simplest way of applying this to the heave motion requires us to assume that the most probable largest heave motion z_{max} in N oscillations (N large) or time, t, can be approximated as

$$z_{max} = \sigma(2 \log N)^{\frac{1}{2}} \qquad (3.59)$$

where

$$\sigma^2 = \int_0^\infty S(\omega) \left| \frac{\eta_3(\omega)}{\zeta_a} \right|^2 d\omega \qquad (3.60)$$

$$N \approx \frac{t}{T_2} \qquad (3.61)$$

This approach is valid for a long-crested sea of given significant wave height $H_{\frac{1}{3}}$, direction and mean wave period T_2. However, by combining these results with results for other wave directions and a scatter diagram for $H_{\frac{1}{3}}$ and T_2 (or modal period T_0) (see Table 2.2) it is possible to obtain long-term statistical values.

It should be noted that the hydrodynamic theory in the mass–force domain implicitly assumes that the incident waves are not disturbed by the platform. In order to evaluate the necessary deck elevation (air gap) it would then be necessary to use the most probable highest wave of a specific return period (for instance, 100 years). On the other hand, we should bear in mind that the theory is based on linearity. To demonstrate that non-linearities have an effect, we will consider an idealized case with regular waves that are periodic in space and time. For a wave of steepness $H/\lambda = 0.1$ the wave elevation at the crest is roughly 20% higher than that predicted by the linear theory (Schwartz, 1974). A steepness of 0.1 is not at all unrealistic for extreme wave situations, which shows that the consequences of non-linearities should be kept in mind. However, at present there are no rational or practical ways of evaluating these effects for an irregular sea, so model tests are needed.

In the wave-interaction domain we have to rely on a three-dimensional source technique. The method is well established as an engineering tool for calculating linear wave-induced motions and loads on large volume structures in regular incident waves without the presence of current. This will be discussed in chapter 4.

HEAVE MOTION OF A SEMI-SUBMERSIBLE

We will analyse semi-submersibles that consist of two pontoons with columns on each floater (see Fig. 3.9). The platform has fore-and-aft symmetry and operates in deep water. The heave motion in beam seas will be studied. The undamped equation of motion in the mass–force domain can be written as

$$(M + A_{33})\frac{d^2\eta_3}{dt^2} + \rho g A_w \eta_3 = F_3(t) \tag{3.62}$$

By following a similar analysis as for the TLP we will find that

$$F_3(t) = \rho g \zeta_a \sin \omega t \, e^{kz_m} \cos(kB/2)\left(A_w e^{k(z_t - z_m)} - k\left(V_p + \frac{A_{33}}{\rho}\right)\right) \tag{3.63}$$

where A_w is the waterplane area of the semi-submersible, z_t and z_m are respectively the z-coordinates of the top and the geometric centre of a pontoon, B is the distance between the centre planes of each pontoon and V_p is the total volume of the pontoons. In deriving equation (3.63) it is assumed that the free-surface elevation at the centre plane of the platform is $\zeta_a \sin \omega t$. By using the fact that $k(z_t - z_m)$ is small we can approximate equation (3.63) by

$$F_3(t) = \zeta_a \sin \omega t \, e^{kz_m} \cos(kB/2)(\rho g A_w - \omega^2(M + A_{33})$$
$$- \rho \omega^2 A_w z_m) \tag{3.64}$$

From (3.62) it follows that

$$\frac{\eta_3}{\zeta_a} = \sin \omega t \, e^{kz_m} \cos(kB/2)\left(1 - \frac{kz_m}{1 - \left(\dfrac{\omega}{\omega_n}\right)^2}\right) \tag{3.65}$$

where

$$\omega_n = \left(\frac{\rho g A_w}{M + A_{33}}\right)^{\frac{1}{2}} \tag{3.66}$$

is the natural circular frequency in heave. The heave motion is either in-phase or $180°$ out-of-phase with the wave elevation at the centre plane of the platform. When $\omega = \omega_n$, the heave motion is infinite. This is unrealistic and is mainly due to neglection of viscous effects. It follows from equation (3.65) that the heave motion is zero when

$$\omega = \frac{\omega_n}{(1 - |z_m| \, \omega_n^2/g)^{\frac{1}{2}}} \tag{3.67}$$

or $\cos(kB/2) = 0$, i.e. $\lambda = 2B/(2n + 1)$ $(n = 0, 1, 2 \ldots)$. The highest period where the theoretical heave value is zero is normally found from equation (3.67). For instance, if $T_n = 22$ s and $|z_m| = 20$ m, we find the cancellation period corresponding to (3.67) is 20.1 sec. Equation (3.65) tells us that the heave response in beam sea in the mass–force domain is mainly a function of the dimensionless parameters ω/ω_n and $\omega_n^2 |z_m|/g$

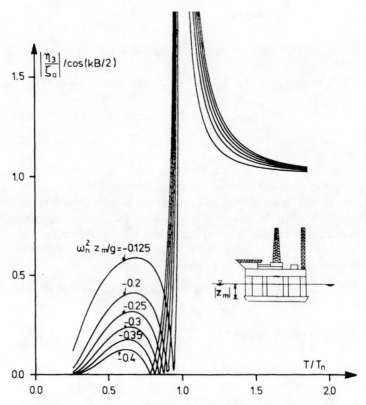

Fig. 3.19. Heave amplitude $|\eta_3|$ in beam sea of the semi-submersible described in the figure. The results presented as a function of main parameters. ($\omega_n = 2\pi/T_n$ = Natural heave frequency (rad/s), $z_m = z$-coordinate of the geometrical centre of the pontoons, B = distance between the centre planes of the pontoons, T = oscillation period, ζ_a = incident wave amplitude.) Infinite water depth.

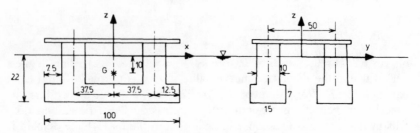

Fig. 3.20. Semi-submersible. Dimensions in metres. (G = centre of gravity).

(see Fig. 3.19). The simplicity of equation (3.65) makes it a good starting point for choosing main platform dimensions that minimize heave motion for extreme weather conditions. In practical semi-submersible design one normally wants to have the heave period above 20 s so that there is seldom any wave energy to excite resonant heave oscillations. Stability and requirements of load carrying capacity governs the requirement on B and the waterplane area A_w. Thus, with selected ω_n, the only parameter that will influence the heave response is the draught z_m. Variation of pontoon geometry will change A_{33} and ω_n (see equation (3.66)).

Example. Heave and pitch motion of a semi-submersible in head sea

Fig. 3.20 shows a semi-submersible with rectangular pontoons and circular vertical columns. The dimensions are in metres. The platform is moored in deep water. Assume the mooring system has no influence on the motions and disregard the effect of damping. Regular sinusoidal waves of period 10 s and amplitude 1 m propagate along the negative x-axis. Calculate the pitch motion and the heave of the centre of gravity G of the platform.

Solution

We disregard the coupling effect from surge. Due to the symmetry of the submerged part of the platform, it is possible to show that the coupling terms A_{35}, A_{53}, C_{35} and C_{53} are equal to zero. We may write the heave equation of the centre of gravity of the semi-submersible as

$$(M + A_{33})\frac{d^2\eta_3}{dt^2} + C_{33}\eta_3 = F_3(t) \tag{3.68}$$

From strip theory it follows that

$$A_{33} = 2LA_{33}^{(2D)}$$

where L is the length and $A_{33}^{(2D)}$ is the two-dimensional added mass in heave for one pontoon. $A_{33}^{(2D)}$ will be set equal to $2.3 \cdot \rho A$ where A is the cross-sectional area of one pontoon. It now follows that

$$M + A_{33} = \rho(2LA + |z_t| A_w) + 2 \cdot 2.3 \, \rho AL = 7.59 \cdot 10^7 \; (\text{kg})$$

The restoring coefficient

$$C_{33} = \rho g A_w = 3.16 \cdot 10^6 \; (\text{kg s}^{-2})$$

The vertical excitation force is obtained by strip theory. We may write

$$F_3 = 2 \int_{-L/2}^{L/2} (\rho A + 2.3 \, \rho A)a_3 \, dx + p_A \frac{A_w}{2} + p_B \frac{A_w}{2}$$

The first term represents the vertical force on the two pontoons assuming no vertical columns. The two last terms are corrections because of the vertical columns. We write the incident wave potential as

$$\phi = \frac{g\zeta_a}{\omega} e^{kz} \cos(\omega t + kx)$$

This means

$$a_3 = -\omega^2 \zeta_a e^{kz_m} \sin(\omega t + kx)$$

where $z_m = -18.5$ m. In the last two terms in the expression for F_3

$$p_A = \rho g \zeta_a e^{kz_1}\sin(\omega t + kx_1), \qquad p_B = \rho g \zeta_a e^{kz_1} \sin(\omega t + kx_2)$$

where $z_1 = -15$ m, $x_1 = -37.5$ m, $x_2 = 37.5$ m. We may now write

$$F_3 = -6.6\rho A \omega^2 \zeta_a e^{kz_m} \frac{2}{k} \sin(kL/2)\sin \omega t$$

$$+ \frac{A_w}{2} \rho g \zeta_a e^{kz_1} 2 \sin \omega t \cos(kx_1)$$

$$= -5.9 \cdot 10^6 \sin \omega t \quad (N)$$

The heave motion of the centre of gravity of the semi-submersible may now be found from equation (3.68) by substituting $\eta_3 = |\eta_3| \sin \omega t$. We find

$$\eta_3 = 0.2 \sin \omega t \quad (m)$$

The pitch equation of the semi-submersible can be written as

$$(I_5 + A_{55}) \frac{d^2\eta_5}{dt^2} + C_{55}\eta_5 = F_5(t) \tag{3.69}$$

The pitch inertia moment I_5 will be set equal to A_{55}. This is not generally true. The pitch added mass A_{55} may be obtained by strip theory and by following the definition of A_{55}, i.e. by studying forced pitch oscillation η_5 and the resulting pitch moment on the semi-submersible. A strip of length dx of the pontoons will be exposed to a vertical acceleration $-x\ddot{\eta}_5$ due to the forced pitch oscillation. This creates a vertical force $A_{33}^{(2D)}x\ddot{\eta}_5 \, dx$ on the strip which again causes a pitch moment about an axis parallel to the y-axis through the centre of gravity. The total contribution to the pitch moment from the pontoons can be written as

$$-2 \int_{-L/2}^{L/2} x^2 A_{33}^{(2D)} \, dx \, \ddot{\eta}_5$$

Let us now study the effect from the columns. A strip of length dz on one of the columns will be exposed to a horizontal acceleration $(z + 10)\ddot{\eta}_5$ due to the forced pitch oscillation. This creates a horizontal force $-A_{11}^{(2D)}(z + 10)\ddot{\eta}_5$, which in turn causes a pitch moment about the y-axis. We may write

$$A_{55} = 2 \int_{-L/2}^{L/2} A_{33}^{(2D)} x^2 \, dx + 4 \int_{-15}^{0} A_{11}^{(2D)} (z + 10)^2 \, dz$$

where $A_{11}^{(2D)} = \rho \pi 10^2 / 4$ and length is measured in metres. We find $A_{55} = 4.1 \cdot 10^{10}$ kg m^2. C_{55} follows from hydrostatic considerations and can be written as

$$C_{55} = \rho g V \overline{GM}_L = 2.8 \cdot 10^9 \text{ kg m}^2 \text{ s}^{-2}$$

The wave exciting pitch moment can be written as

$$F_5 = \int_{-L/2}^{L/2} 6.6 \rho A \omega^2 \zeta_a e^{kz_m} \sin(\omega t + kx) x \, dx$$

$$- x_1 \frac{A_w}{2} \rho g \zeta_a e^{kz_1} \sin(\omega t + kx_1)$$

$$- x_2 \frac{A_w}{2} \rho g \zeta_a e^{kz_1} \sin(\omega t + kx_2)$$

$$- 2 \int_{-15}^{0} (z + 10)(0.25 \rho A_w$$

$$+ A_{11}^{(2D)}) \omega^2 \zeta_a e^{kz} \cos(\omega t + kx_1) \, dz$$

$$- 2 \int_{-15}^{0} (z + 10)(0.25 \rho A_w$$

$$+ A_{11}^{(2D)}) \omega^2 \zeta_a e^{kz} \cos(\omega t + kx_2) \, dz$$

The two last terms are contribution to the wave exciting pitch moment from the horizontal forces on the columns. If we neglect the last two terms, we will find $F_5 = 2.3 \cdot 10^8 \cos \omega t$(Nm).

By substituting $\eta_5 = |\eta_5| \cos \omega t$ in equation (3.69) it now follows that the pitch motion is

$$\eta_5 = -0.008 \cos \omega t \text{(rad)}$$

MINIMALIZATION OF VERTICAL SHIP MOTIONS

We now discuss which parameters influence the heave and pitch of the ship. In order to do this we choose a simplified mathematical model and consider first a ship of constant cross-section in head sea at zero Froude number. We do not recommend the use of this model for accurate calculations, the intention here being to show general trends. We will use strip theory and a long wavelength approximation for the wave excitation loads (see equation (3.44)). The latter implies that the vertical wave excitation load on a strip is written as the sum of the vertical Froude–Kriloff force and a vertical diffraction force consisting of two parts. The first part is the cross-sectional added mass in heave multiplied by the vertical incident wave acceleration at a representative point of the strip. The second part is the cross-sectional damping in heave multiplied by the vertical incident wave velocity at a representative point of the strip. It is assumed that the cross-sectional dimension is small relative to the wavelength. By following this approach we will find that the amplitude of vertical wave excitation force in head seas $|F_3|_{head\,sea}$ can be related to the amplitude of the vertical wave excitation force in beam seas $|F_3|_{beam\,sea}$ by the formula

$$|F_3|_{head\,sea} = |F_3|_{beam\,sea} \cdot \frac{2}{kL} \left| \sin\left(\frac{kL}{2}\right) \right| \tag{3.70}$$

Here L is the length of the ship and $k = 2\pi/\lambda$ is the wavenumber of the incident waves. Equation (3.70) shows the way that the wave loads along the ship in head seas may counteract each other and cancel the total wave excitation loads on the ship. Equation (3.70) implies that the heave amplitude $|\eta_3|_{head\,sea}$ in head seas can be related to the heave amplitude in beam seas $|\eta_3|_{beam\,sea}$ by the formula

$$|\eta_3|_{head\,sea} = |\eta_3|_{beam\,sea} \cdot \frac{2}{kL} \left| \sin\left(\frac{kL}{2}\right) \right| \tag{3.71}$$

Similarly we can analyse the amplitude of the pitch angle $|\eta_5|$. If we assume constant mass distribution along the length of the ship we can write

$$|\eta_5|_{head\,sea} = |\eta_3|_{beam\,sea} \cdot \frac{12}{L} \left| -\frac{1}{kL} \cos\left(\frac{kL}{2}\right) \right.$$
$$\left. + \frac{2}{L^2 k^2} \sin\left(\frac{kL}{2}\right) \right| \tag{3.72}$$

The analysis of the ship at infinite Froude number becomes more complicated. In this context we will only incorporate the important effect of the frequency of encounter ω_e which is written as

$$\omega_e = \omega_0 + \frac{\omega_0^2}{g} U \tag{3.73}$$

Here $\omega_0^2/g = 2\pi/\lambda = k$ and U is the forward speed of the ship.

Equation (3.73) is valid for head seas. If we assume the same natural frequency at forward speed as in zero speed, equation (3.73) tells us that it is a larger wavelength of the incident waves that creates resonance in heave and pitch at forward speed than at zero speed. This is illustrated in Fig. 3.21. For a real ship form at zero forward speed the cancellation effect on the heave wave excitation force may be dominant around the natural frequency for heave and pitch. Due to the effect of the frequency of encounter, the cancellation effect on the heave and pitch wave excitation loads will be less pronounced at forward speed. The consequences are increased heave and pitch at resonance within certain limits of increasing Froude number.

From equations (3.71) and (3.72) we see that it is relevant to consider heave in beam seas when we discuss heave and pitch in a head sea. We will use heave at resonance in beam seas as a basis for this discussion. If we use equation (3.45) for the wave excitation loads we find

$$|\eta_3|_{\text{beam sea}} = \zeta_a \frac{g}{\omega_{n3}^{3/2}} \left[\frac{\rho}{B_{33}^{(2D)}} \right]^{\frac{1}{2}} \quad \text{at heave resonance} \quad (3.74)$$

Here ζ_a is the undisturbed wave amplitude, ω_{n3} the natural circular frequency of oscillation in heave and $B_{33}^{(2D)}$ is the cross-sectional two-dimensional damping in heave. $B_{33}^{(2D)}$ is a function of ω_{n3}. In two-dimensional problems for ship cross-sections either source techniques or conformal mapping techniques are most commonly used to calculate added mass and damping in heave. Lewis-form technique is an approximate conformal mapping technique. We know that the Lewis-form technique is a satisfactory method to find the added mass and damping coefficients of most of the cross-sectional forms of ships. This implies that the beam–draught ratio B/D and the sectional area

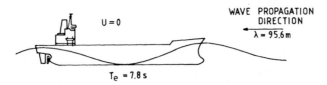

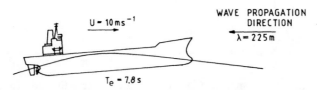

Fig. 3.21. Illustration to show that the frequency of encounter corresponds to different wavelengths for zero and non-zero forward speed of a ship in head sea. (T_e = encounter period, λ = wavelength, U = forward speed.)

coefficient $\sigma = A/(B \cdot D)$ are sufficient hull parameters to determine the added mass and damping coefficients. However, this is not true for bulb sections for instance. By using results from Tasai (1959) for the added mass and damping coefficient in heave we can evaluate equation (3.74). The natural frequency ω_{n3} in equation (3.74) depends on the added mass coefficient. This means ω_{n3} will not be the same for different cross-sectional forms of our ship. The results for heave at resonance in beam sea as a function of beam–draught ratio B/D and sectional area coefficient $\sigma = A/(B \cdot D) = C_B$ are plotted in Fig. 3.22. We can conclude that the heave motion decreases with increasing beam–draught ratio and increases with increasing sectional area coefficient. In reality the sectional area coefficient corresponds to the block-coefficient for our particular choice of ship.

From the discussion above one may be tempted to conclude that as long as the cancellation effects of the wave loads along the ship are not pronounced, the heave and pitch motion in head sea at resonance decrease with increasing beam–draught ratio and decreasing block-coefficient. On the other hand, if the cancellation effects are pronounced we should be careful in making such a statement. The reason is that the natural periods in heave and pitch are dependent on the hull parameters and we will get different cancellation effects at the heave and pitch natural periods of the different hulls. In addition the heave and pitch

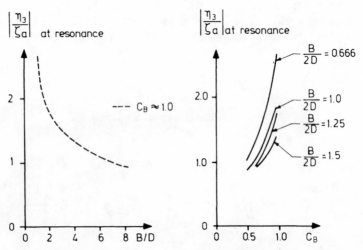

Fig. 3.22. Heave amplitude $|\eta_3|$ at resonance of a ship in beam sea as a function of beam–draught ratio and block coefficient. The ship has a constant cross-section along its length. (ζ_a = incident wave amplitude, B = beam, D = draught, C_B = block coefficient.) Infinite water depth.

motion increase with increasing Froude number within certain limits. The conclusions above are based on discussions of transfer functions.

If we combine the transfer functions with a sea spectrum it is obvious that the root mean square values of the vertical motions will depend on the ship length L. In addition it is expected that the vertical ship motions as a general trend decrease with increasing ship length. We do not say that this always occurs. This depends on the peak period of the wave spectrum and the ship length.

The vertical motions may depend on hull parameters other than C_B, B/D and L. Since the waterplane area coefficient C_W can be thought of as representing the local beam–draught ratio along the ship we may be tempted to conclude that increasing C_W and keeping draught constant means decreasing heave and pitch motions. Based on our simplified theoretical model it would be speculative to include any new parameters, but we should be aware that there is no coupling between heave and pitch in our simplified model. One reason is that we have selected a ship model with fore- and aft-symmetry. It is known that coupling effects between heave and pitch are of some importance. It would therefore be logical to introduce a hull parameter characterizing the fore- and aft-symmetry of the hull, for instance LCB–LCF where LCB and LCF are respectively the longitudinal centre of buoyancy and floatation.

If we compare our conclusions with Schmitke & Murdey's (1980) conclusions from a systematic investigation of frigate hulls in different sea states, we are in agreement about the influence of L, B/D, C_W and C_B. Gerritsma et al. (1974) did a series of tests and calculations for regular waves where the length, draught and block coefficient were kept constant. The beam and the Froude number were varied. Their results seem to be in agreement with our conclusions that heave and pitch at resonance decrease with increasing beam–draught ratio, and increase with increasing Froude number.

It has been reported in the literature that the vertical ship motions depend on the bulb, transom stern, and the pitch radius of gyration. Further, ships with U-form seem to have different heave and pitch motions than ships with V-form. Our simplified theoretical model is not able to explain this.

We should note that heave and pitch motions are not sensitive to small changes in hull form. However, we should be careful in generalizing the findings to relative vertical motions between the ship and the waves. The phases of heave and pitch motions are then important in the evaluation.

ROLL STABILIZATION

Rolling of a ship can be a problem and roll amplitudes of 30 and 40° have occurred on ships. Large roll angles make it difficult for the crew to do

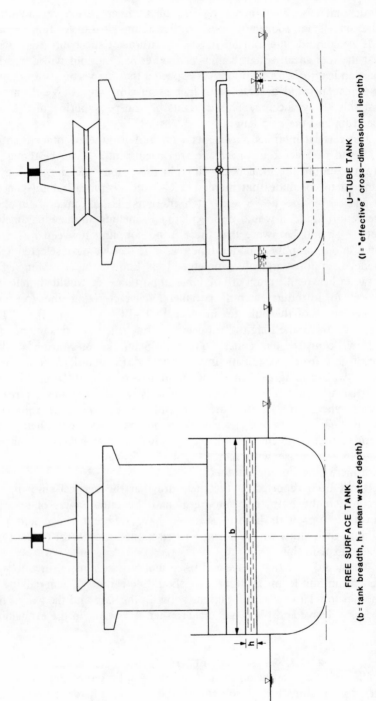

FREE SURFACE TANK
(b = tank breadth, h = mean water depth)

U-TUBE TANK
(l = "effective" cross-dimensional length)

Fig. 3.23. Passive anti-roll tank stabilisers.

their job. Many ships have therefore been equipped with roll stabilization equipment, to avoid a strong resonance effect in roll.

The importance of bilge keels to damp roll motion is well known. The bilge keel damping is due both to the resistance of the bilge keel through the water and the effect of bilge keels on the pressure distribution around the body. Empirical formulas for the bilge keel damping can be found in Kato (1966) and Ikeda et al. (1976, 1977b). In this context we will concentrate on describing passive anti-roll tanks as means to dampen roll motions.

Passive anti-roll tanks

Passive anti-rolling tanks have been installed on many ships. Principles for the free-surface tank and the U-tube tanks are shown in Fig. 3.23. By changing the water level it is possible to change the resonance frequency of the free-surface tank. This type of rolling tank is therefore especially well suited for ships operating with a broad spectrum of metacentric heights.

A major problem with anti-roll tanks is saturation. When the ship motions become large, the water motion may hit the tank top. This reduces the effect of the system.

In order for the anti-rolling tank to work satisfactorily, it is necessary for the natural period of the water motion in the tank to be close to the natural period of roll motion. Some people choose to set these two eigenfrequencies to be equal, while others prefer to set the eigenfrequency for the passive tank to be 6–10% higher than the eigenfrequency in rolling. The tank's damping effect will be improved by an increased ratio between the tank's and the ship's metacentric height. Typical passive tanks have a ratio of $\delta \overline{GM}_T / \overline{GM}_T$ between 0.15 to 0.3, where $\delta \overline{GM}_T$ means a decrease in the ship's metacentric height due to anti-rolling tanks. By means of the tank's natural period and the metacentric height ratio $\delta \overline{GM}_T / \overline{GM}_T$ it is possible to determine the tank dimensions.

The highest natural period of a rectangular free surface tank of infinite length can be written as

$$T_N = 2\pi \left/ \left(\frac{g\pi}{b} \tanh\left(\frac{\pi h}{b} \right) \right)^{\frac{1}{2}} \right. \tag{3.75}$$

where h is the water depth in the tank and b is the breadth of the tank (see equation (2.42)). This is the natural period that is most likely to be in the vicinity of the roll natural period.

The waterdepth h will in practice be small compared to b. We can then write

$$T_N \simeq \frac{2b}{(gh)^{\frac{1}{2}}} \tag{3.76}$$

This shows in a simple way how the tank breadth and tank depth influence the natural tank period.

For a U-tube tank of constant cross-section A we can write the natural period as

$$T_N = 2\pi \left(\frac{l}{2g} \right)^{\frac{1}{2}} \tag{3.77}$$

where l is a length dimension of the water, defined so that $l \cdot A$ is the fluid volume (see exercise 3.6(d)).

The reason why the natural period should be selected close to the natural period of roll motion is that the tank moment is very closely in phase with the roll velocity and has maximum damping effect when the tank is forced to oscillate in roll at the natural period of the tank motion.

This is also illustrated in Fig. 3.24. At the resonance of the fluid motion a hydraulic jump is formed. The position of the jump is close to the mid-point of the tank when the roll velocity is at a maximum. The hydrodynamic moment due to the fluid motion inside the tank can be calculated by noting that the fluid pressure is static below the instantaneous free-surface position. We note that this causes a moment against the roll velocity. The maximum moment due to the fluid motion occurs when the hydraulic jump is at the mid-point, i.e. when the roll velocity is at a maximum.

We will now show how to choose the main dimensions h, b and L of a free-surface tank in a pre-design phase. L is the length of the tank. We choose b to be the cross-sectional beam of the ship where the tank is going to be installed. The water depth h is chosen by requiring that

$$(T_N)_{\text{tank}} = (T_N)_{\text{roll}}$$

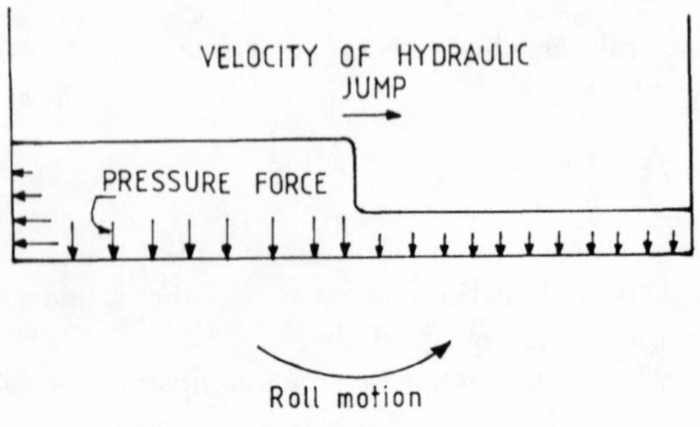

Fig. 3.24. Simplified picture of fluid motion in a shallow water tank at resonance condition shown at the time instant when the roll angle is zero and the roll velocity is maximum.

This means (see equation (3.76)) that the water depth

$$h = \frac{1}{g} \frac{4b^2}{(T_\mathrm{N})_\mathrm{roll}^2} \tag{3.78}$$

The length L of the tank is chosen so that $\overline{\delta GM_T}/\overline{GM_T}$ is between 0.15 and 0.3. We must obviously have the hydrostatic stability requirements in mind when we do that. In order to assess the effect of the anti-rolling tanks accurate model tests are needed.

EXERCISES

3.1 Motions and sea loads on a barge (Fig. 3.25)
Suppose a barge is a box-shaped floating body with length $L = 200$ m, beam $B = 30$ m and draught $D = 15$ m. Assume constant mass density, and that the barge has zero forward speed. Assume regular sinusoidal waves that propagate along the negative x-axis, and use a wavelength λ of 300 m and wave height of 20 m. Define the velocity potential of the incident waves as

$$\phi = \frac{g\zeta_a}{\omega} e^{kz} \cos(\omega t + kx)$$

(a) Show that the vertical excitation force can be written as

$$F_3 = \{\rho g \zeta_a B e^{-kD} - g A_{33}{}^{(2\mathrm{D})} k \zeta_a e^{-kD/2}\} \frac{2}{k} \sin\left(\frac{kL}{2}\right) \sin \omega t$$

by means of long wavelength formulas and strip theory.
Discuss applicability of this approximation and what happens when $\omega \to 0$.

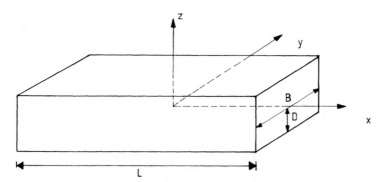

Fig. 3.25. Definition of parameters and coordinate system used in exercise 3.1 for the analysis of a barge.

(b) Use $A_{33}^{(2D)} = 0.8\rho BD$, neglect damping and show that the heave motion of the centre of gravity is

$$\eta_3 = 4.9 \sin \omega t \quad (m)$$

Discuss the heave motion relative to the incident waves with respect to both amplitude and phase.

(c) Based on the assumptions of (a) and (b), show that the pitch excitation moment is

$$F_5 = \frac{\rho g \zeta_a 2B}{k} \left\{ \frac{L}{2} \cos\left(\frac{kL}{2}\right) - \frac{1}{k} \sin\left(\frac{kL}{2}\right) \right\}$$
$$\times \cos \omega t \{e^{-kD} - 0.8kDe^{-kD/2}\}$$

(d) Show that the pitch motion can be written as

$$\eta_5 = -0.15 \cos \omega t \text{ (rad)}$$

Discuss the pitch motion relative to the incident wave slope with respect to both amplitude and phase.

(e) Show that the amplitude of the vertical acceleration in the bow $(x = L/2)$ is 0.34 g.

(f) Show that the relative vertical motion between the barge and the incident waves in the bow is

$$9.9 \sin \omega t + 6.7 \cos \omega t \quad (m)$$

Will the barge bottom go out of the water? What must the freeboard be to avoid water on deck?

(g) Find the vertical dynamic bending moment midships. Show that the contribution from wave excitation loads, restoring loads, added moment loads and inertia moment loads can be written as respectively

$$I_1 = \frac{\rho g \zeta_a B}{k} \{e^{-kD} - 0.8kDe^{-kD/2}\}$$
$$\times \left\{ \sin \omega t \left[\frac{1}{k} - \frac{1}{k} \cos\left(\frac{kL}{2}\right) - \frac{L}{2} \sin\left(\frac{kL}{2}\right) \right] \right.$$
$$\left. + \cos \omega t \left[\frac{L}{2} \cos\left(\frac{kL}{2}\right) - \frac{1}{k} \sin\left(\frac{kL}{2}\right) \right] \right\}$$

$$I_2 = \frac{1}{2}\left(\frac{L}{2}\right)^2 \rho g B \eta_3 - \frac{1}{3}\left(\frac{L}{2}\right)^3 B \rho g \eta_5$$

$$I_3 = \frac{1}{2}\left(\frac{L}{2}\right)^2 A_{33}^{(2D)} \frac{d^2\eta_3}{dt^2} - \frac{1}{3}\left(\frac{L}{2}\right)^3 A_{33}^{(2D)} \frac{d^2\eta_5}{dt^2}$$

$$I_4 = -\left\{ \frac{1}{2}\left(\frac{L}{2}\right)^2 \rho BD \frac{d^2\eta_3}{dt^2} - \frac{1}{3}\left(\frac{L}{2}\right)^3 \rho BD \frac{d^2\eta_5}{dt^2} \right\}$$

The wave bending moment midships is given as

$$BM = I_1 + I_2 + I_3 - I_4 \approx 2.1 \cdot 10^9 \sin \omega t \quad (Nm)$$

3.2 Resonant heave motion of surface effect ships (SES)

SES (see Fig. 3.26) is an air-cushion supported vehicle where the air cushions are enclosed on the sides by rigid sidewalls and on the bow and stern by compliant seals. Resonance heave motion may cause excessive vertical accelerations which are unpleasant and limit operations on board a SES in cushion-borne condition.

(a) The resonance period of the heave motion of the centre of gravity of a cushion-borne SES is much lower than for monohull ships with comparable length. The natural period for heave of the centre of gravity of a cushion-borne SES can be approximated by

$$T_{n3} = 2\pi \left[\frac{Mh_b}{\gamma A_b(p_0 + p_a)} \right]^{\frac{1}{2}} \approx 2\pi \left[\frac{h_b}{\gamma g(1 + p_a/p_0)} \right]^{\frac{1}{2}} \qquad (3.79)$$

where γ = specific heat ratio of air $(=1.4)$

g = gravitational constant

p_a = atmospheric pressure

p_0 = difference between cushion pressure and atmospheric pressure

h_b = cushion plenum height

A_b = cushion area

M = mass of the SES

This is also the natural period for the cushion and arises from the compressibility of the air in the cushion.

We shall show how we can derive equation (3.79). We will neglect hydrodynamic forces on the sidewalls and the seals and assume an adiabatic pressure–density relation for the air. Show that for small dynamic pressure variations Δp in the air cushion we may write the following linearized pressure–density relation

$$\rho = \rho_a[1 + \mu/[\gamma(1 + p_a/p_0)]]$$

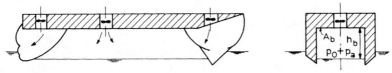

Fig. 3.26. Surface Effect Ship (SES).

where $\mu = \Delta p/p_0$ and ρ_a is the air density at the pressure $p_a + p_0$. (Hint: $p = \alpha \rho^\gamma$ where α is constant).

We will assume the air cushion is rectangular-box-shaped and neglect air leakage and effect of fan flow. Show that the linearized continuity equation for the air in the cushion can be written as

$$A_b \frac{d\eta_3}{dt} + \frac{dV_W}{dt} + \frac{A_b h_b}{\gamma(1 + p_a/p_0)} \frac{d\mu}{dt} = 0$$

where V_W is the change in the air cushion volume due to the waves. (Hint: Use that $d(\rho\Omega)/dt = 0$ where Ω is air cushion volume).

Show from Newton's law that

$$M \frac{d^2\eta_3}{dt^2} - p_0 A_b \mu = 0$$

Show equation (3.79) by neglecting the forcing function in the continuity equation and by assuming the solutions for heave and dynamic pressure to be of the form $e^{i\sigma t}$.

(b) Consider a 200 tonne ship with a cushion length 40 m, cushion beam 10 m and a cushion plenum height 2 m. Show that $T_{n3} = 0.5$ s if the effect of the buoyancy of the rigid sidewalls is neglected.

(c) Show that the ratio between T_{n3} in model and in full scale is proportional to L_M/L if model tests are done with a model at length scale ratio L_M/L (L_M = model scale length, L = full scale length). What are the consequences of this when model tests of heave accelerations are done?

(d) Consider a sea state with a modal period $T_0 = 2.5$ s. Assume long-crested head sea waves and assume for simplicity one regular wave system with period $T_0 = 2.5$ s. What vessel speed causes heave resonance?

(e) Ride control systems are used to damp resonance heave oscillations. By examining equation (3.79) discuss what additional possibilities there are to reduce the effect of resonance heave oscillations in an operational situation.

(f) Assume the surface wave elevation is given by $\zeta_a \sin(\omega_e t + kx)$. Show that the wave volume pumping $dV_W/dt = -A_b \zeta_a \omega_e \cos(\omega_e t) \sin(\pi L/\lambda)/(\pi L/\lambda)$. L = length of the air cushion.

3.3 Vertical motions of hydrofoil catamarans

Fig. 3.27 shows an example on a hydrofoil-catamaran with four foils. At design speed most of the weight of the vessel is carried by the hydrofoils.

Depending on the design the buoyancy of the catamaran hulls may be 0–30% of the weight of the vessel at maximum speed.

Consider incident regular head sea waves in deep water. Define the velocity potential of the incident waves as

$$\phi = \frac{g\zeta_a}{\omega_0} e^{kz} \cos(\omega_e t - kx)$$

The coordinate system is defined in Fig. 3.2. Assume the wavelength is large relative to the cross-dimensions of the foil. Neglect the hydrodynamic forces on the catamaran hulls and use strip theory for the foils. Assume quasi-steady flow when the lift forces on the foils are analysed. The effect of the free-surface can be neglected. The lift force F_L per unit length on a two-dimensional section of the foil will be expressed as

$$F_L = \frac{\rho}{2} C_L l U^2$$

where l is the chord-length of the foil and C_L is the lift coefficient. C_L depends on the foil characteristics and the angle of attack α of the incident flow velocity relative to the foil.

(a) Explain that the linear dynamic lift force per unit length on the foil can be approximated by

$$F_{LD} = \frac{\rho}{2} \frac{dC_L}{d\alpha}\bigg|_{\alpha=\alpha_0} lU\Big(\omega_0 \zeta_a e^{kz_i} \cos(\omega_e t - kx_i)$$

$$-\frac{d\eta_3}{dt} + x_i \frac{d\eta_5}{dt} + U\eta_5 \Big)$$

where (x_i, y, z_i) is the geometric centre of the foil and $\alpha = \alpha_0$ is the angle of attack of the incident flow in still water.

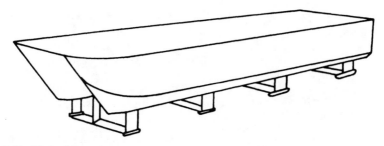

Fig. 3.27. Hydrofoil-catamaran.

(b) There are additional vertical hydrodynamic forces per unit length on the foil that can be written approximately as

$$F_{3A} = \rho\pi\left(\frac{l}{2}\right)^2\left(-\omega_0\omega_e\zeta_a e^{kz_i}\sin(\omega_e t - kx_i) - \frac{d^2\eta_3}{dt^2} + x_i\frac{d^2\eta_5}{dt^2}\right)$$

Explain this when $U = 0$. (Hint: Explain the contribution from the first term in the brackets in a similar way as we did in explaining the diffraction part of the mass term in Morison's equation (see equation (3.34))).

(c) Consider a hydrofoil-catamaran like the one in Fig. 3.27 but with two foils. The centre of gravity is midships and the longitudinal position of the foils are $x_i = \pm L/2$. The foils are equal and the span is s. Show that the added mass and damping coefficients in heave and pitch for the vessel can be written as

$$A_{33} = 2\pi\rho\left(\frac{l}{2}\right)^2 s$$

$$B_{33} = \rho\frac{dC_L}{d\alpha}\bigg|_{\alpha=\alpha_0} lsU$$

$$A_{35} = A_{53} = B_{35} = B_{53} = 0$$

$$A_{55} = A_{33}(L/2)^2$$

$$B_{55} = B_{33}(L/2)^2$$

Show that the foils cause the following non-zero restoring coefficient

$$C_{35} = -\rho\frac{dC_L}{d\alpha}\bigg|_{\alpha=\alpha_0} lsU^2$$

Show that the vertical excitation force F_3 and pitch excitation moment F_5 can be written as

$$F_3 = -2\rho\pi\left(\frac{l}{2}\right)^2 s\omega_0\omega_e\zeta_a e^{kz_i}\cos\left(k\frac{L}{2}\right)\sin\omega_e t$$

$$+ \rho\frac{dC_L}{d\alpha}\bigg|_{\alpha=\alpha_0} lsU\omega_0\zeta_a e^{kz_i}\cos\left(k\frac{L}{2}\right)\cos\omega_e t$$

$$F_5 = -\rho\pi\left(\frac{l}{2}\right)^2 s\omega_0\omega_e\zeta_a e^{kz_i}L\sin\left(k\frac{L}{2}\right)\cos\omega_e t$$

$$- \rho\frac{1}{2}\frac{dC_L}{d\alpha}\bigg|_{\alpha=\alpha_0} lsUL\omega_0\zeta_a e^{kz_i}\sin\left(k\frac{L}{2}\right)\sin\omega_e t$$

(d) Consider an example of the case presented in (c). The vessel mass $M = 1.3 \cdot 10^5$ kg, the pitch radius of gyration $r_{55} = 8.9$ m, the length L of the vessel is 30 m, the chord length $l = 1$ m, the foil span s is 5 m, $\mathrm{d}C_L/\mathrm{d}\alpha = 5.5$ and the ship speed $U = 25$ m s^{-1}. Assume there is some buoyancy effect of the hulls and include the dynamic effect of the hulls only as hydrostatic restoring terms. The total waterplane area is 30 m^2. C_{35} and C_{53} due to the hulls are zero and $C_{55}/C_{33} = 60$ m^2.

Consider the uncoupled equations for heave and pitch. What are the undamped natural periods T_{n3} and T_{n5} for heave and pitch? What are the ratios ζ of the damping relative to the critical damping? (Hint: The critical damping in heave is $2(M + A_{33})2\pi/T_{n3}$.) (Answers: $T_{n3} = 4.3$ s, $T_{n5} = 5.2$ s, ζ for heave: 1.7, ζ for pitch: 5.3.)

3.4 Sway and roll motion of a buoy

Consider a buoy in regular incident waves in deep water. The buoy is a cylinder with circular cross-section and the incident waves are propagating along the positive y-axis (see Fig. 3.28). The wavelength λ is assumed to be larger than 5 times the diameter so that the buoy does not generate any waves of significance. Linear potential theory can be assumed. In the figure, B means the centre of buoyancy and G the centre of gravity. The coupled equations for sway and roll can formally be written

$$(M + A_{22})\frac{\mathrm{d}^2\eta_2}{\mathrm{d}t^2} + A_{24}\frac{\mathrm{d}^2\eta_4}{\mathrm{d}t^2} = F_2 \cos \omega t$$

$$A_{42}\frac{\mathrm{d}^2\eta_2}{\mathrm{d}t^2} + (I_4 + A_{44})\frac{\mathrm{d}^2\eta_4}{\mathrm{d}t^2} + \rho g V \overline{GM}_T \eta_4 = F_4 \cos \omega t$$

where η_2 is the sway motion of the centre of gravity. The moments are referred to the x-axis through the centre of gravity. I_4 is the moment of inertia in roll. The velocity potential for the incident waves is written as

$$\phi = \frac{g\zeta_a}{\omega} e^{kz} \cos(\omega t - ky)$$

(a) Show by strip theory that

$$A_{22} = \rho A d, \quad A_{24} = -\rho A d \overline{BG},$$

$$A_{42} = -\rho A d \overline{BG}, \quad A_{44} = \rho A\left(\frac{\mathrm{d}^3}{12} + d\overline{BG}^2\right)$$

$$F_2 = 2\rho g A \zeta_a (1 - e^{-kd}), \quad F_4 = -2\rho g A \zeta_a(C + D e^{-kd})$$

where

$$C = \left(\frac{d}{2} + \overline{BG}\right) - \frac{1}{k}, \qquad D = \frac{d}{2} - \overline{BG} + \frac{1}{k}$$

(b) Show that the natural circular frequency for the coupled oscillation in roll and sway is

$$\omega = \left(\frac{\rho g V \overline{GM}_T}{(I_4 + A_{44}) - \frac{1}{2}M\overline{BG}^2}\right)^{\frac{1}{2}}$$

(c) Assume a diameter $D = 2$ m, the draught $d = 10$ m, $\overline{BG} = 1$ m and the height of the buoy $L = 14$ m. The wavelength $\lambda = 20$ m and the wave height is 0.5 m. The roll inertia moment with respect to the axis through the centre of gravity is $I_4 = 0.15ML^2$.

(i) Show that the natural period for coupled roll and sway motion is 12.3 s.

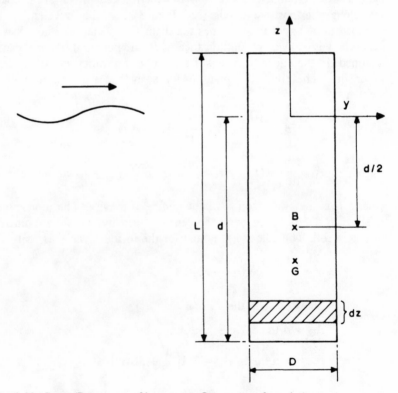

Fig. 3.28. Buoy (B = centre of buoyancy, G = centre of gravity).

(*ii*) Show that the roll motion can be written as

$$\eta_4 = 0.012 \cos \omega t \ (\text{rad})$$

and the sway motion of the centre of gravity is

$$\eta_2 = -0.07 \cos \omega t \ (\text{m})$$

Discuss how the buoy sways and rolls relative to the incident waves.

(*iii*) Is there any point on the buoy where the horizontal motion is always equal to zero?

(*iv*) Find the horizontal acceleration of the top of the buoy (Answer: $0.59 \cos \omega t \ (\text{m s}^{-2})$).

3.5 Roll motions of ships

We will assume a beam sea and use strip theory. For simplicity we will assume the roll motion is uncoupled from other motions. The uncoupled linear roll equation in regular waves can be written as

$$(I_4 + A_{44})\frac{d^2\eta_4}{dt^2} + B_{44}\frac{d\eta_4}{dt} + C_{44}\eta_4 = F_4 \sin \omega t \tag{3.80}$$

(*a*) Give a short explanation of the terms in equation (3.80) and be precise about what moment axis is used.

(*b*) Assume that the damping term is due to wave radiation. Derive a relationship between the far-field radiated waves due to forced roll motion and B_{44}.

(*c*) The steady-state solution of equation (3.80) can be written as

$$\eta_4 = \frac{F_4 \sin(\omega t + \theta_1)}{C_{44}\left[\left(1 - \left(\frac{\omega}{\omega_n}\right)^2\right)^2 + \left(2\zeta\frac{\omega}{\omega_n}\right)^2\right]^{\frac{1}{2}}} \tag{3.81}$$

where

$$\omega_n = \left(\frac{C_{44}}{I_4 + A_{44}}\right)^{\frac{1}{2}}$$

$$\zeta = \frac{B_{44}}{2(I_4 + A_{44})\omega_n} \quad \begin{array}{l}\text{(fraction between damping and critical} \\ \text{damping } 2(I_4 + A_{44})\omega_n)\end{array}$$

$$\theta_1 = \arctan\left(\frac{-2\zeta\dfrac{\omega}{\omega_n}}{1 - \dfrac{\omega^2}{\omega_n^2}}\right)$$

Assume the ship has a constant cross-section. Express the roll amplitude at resonance frequency ω_n in terms of only $B_{44}^{(2D)}$, ω_n, ζ_a, ρ and g.

(d) In practice viscous roll damping is important. A modified roll equation can then be written

$$(I_4 + A_{44})\frac{d^2\eta_4}{dt^2} + B_{44}\frac{d\eta_4}{dt} + B_V\frac{d\eta_4}{dt}\left|\frac{d\eta_4}{dt}\right| + C_{44}\eta_4 = F_4 \sin \omega t$$

(3.82)

When equation (3.82) is solved, it is usual to linearize the non-linear term, by introducing a linear term $K\,d\eta_4/dt$ so that the work done over one period is the same for this term and the non-linear term. Show that

$$K = B_V\,|\eta_4|\,\frac{8\omega}{3\pi}$$

(3.83)

This is called the method of equivalent linearization.

(e) If the effect of bilge keels is accounted for, this is only done in the damping term. Give an estimate what the effect of bilge-keels would be on the mass term by considering a ship with constant circular cross-section and with moment axis coinciding with the cylinder axis. Assume that the bilge keel has a radial direction, zero thickness and a maximum breadth of 2% of the beam of the ship. Neglect free-surface effects and the curvature of the hull in the analysis of the bilge keel.

Derive an expression for the roll damping of the bilge keels for the same situation. Use the fact that the normal force F_n on the bilge keel can be written as $F_n = -0.5\rho C_D A u\,|u|$ where u is the velocity normal to the bilge keel as if the bilge keel were not there. Further, A is the frontal area of the bilge keel and C_D is the proper drag coefficient, which can be set equal to $8.0\,KC^{-\frac{1}{3}}$ where $KC = u_{max}T/(2b)$ (u_{max} = maximum value of u, T = oscillation period, b is the bilge keel breadth).

(f) In an irregular sea a different linearization technique has to be used. We cannot use equation (3.83). One can show (see Price & Bishop, 1974) that

$$K = 2\left(\frac{2}{\pi}\right)^{\frac{1}{2}}B_V\sigma_{\dot{\eta}_4}$$

(3.84)

where $\sigma_{\dot{\eta}_4}$ is the standard deviation of the roll velocity. Describe briefly a procedure to find $\sigma_{\dot{\eta}_4}$ and σ_{η_4}.

(g) In practice, coupling between roll and sway is important in the case presented above. Suppose the coupling effect is accounted for. Is it possible to find a horizontal axis on the ship where there is no horizontal motion? (This axis is sometimes referred to as the roll axis.)

(h) Assume that the sway, roll and yaw motions are known. Derive an expression for the lateral acceleration in a body-fixed coordinate system.

3.6 Fluid motion in a moonpool

Consider a moonpool as shown in Fig. 3.29. Resonance oscillations may occur in the moonpool. We will analyse this by using a linear theory. It is assumed that the ship motions are known.

The figure represents a longitudinal cross-section of the ship. The moonpool is assumed to have a constant horizontal circular cross-section with diameter D. We will assume the water motion does not vary across the moonpool. This means there is a constant vertical velocity $d\eta/dt$ in the moonpool, where η is the free surface elevation in the moonpool.

(a) Show by differentiating Bernoulli's equation (see equation (2.4)) that the vertical pressure gradient $\partial p/\partial z$ can be related to the vertical fluid acceleration by

$$\frac{d^2\eta}{dt^2} = -\frac{1}{\rho}\frac{\partial p}{\partial z} - g \tag{3.85}$$

(b) Integrate equation (3.85) from $z = -h$ to $z = \eta$ and show that the following linearized equation follows

$$\frac{d^2\eta}{dt^2} + \frac{g}{h}\eta = -\frac{1}{h}\frac{\partial \phi}{\partial t}\bigg|_{z=-h} \tag{3.86}$$

(c) Equation (3.86) is like a mass–spring system without damping. The right hand side represents the exciting 'force'. Show that the natural period T_n of the system is

$$T_n = 2\pi\sqrt{\frac{h}{g}} \tag{3.87}$$

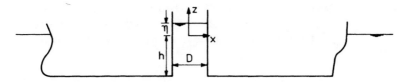

Fig. 3.29. Moonpool dimensions.

(d) Equation (3.87) resembles the natural period for a U-tube tank (see equation (3.77)). Consider a U-tube tank as shown in Fig. 3.23 with constant cross-section. Follow the procedure outlined in (a) and (b) and show that equation (3.77) holds.

(e) We see that equation (3.86) leaves us with an infinite response if the wave period T is equal to T_n. It is likely that T will be close to T_n. For instance if $h = 10$ m we see that $T_n \approx 6$ s. In order to predict realistic values at resonance we need to introduce damping terms. We will assume this is due to viscous effects and write

$$\frac{d^2\eta}{dt^2} + B_D \frac{d\eta}{dt} \left| \frac{d\eta}{dt} \right| + \frac{g}{h}\eta = -\frac{1}{h}\frac{\partial\phi}{\partial t}\bigg|_{z=-h} \tag{3.88}$$

Let us consider an irregular sea described by a sea spectrum $S(\omega)$. In this case the linearization has to be done in a stochastic sense (see for instance (3.84) in the discussion on non-linear roll damping). The result is

$$K = 2\left(\frac{2}{\pi}\right)^{\frac{1}{2}} B_D \sigma_{\dot{\eta}} \tag{3.89}$$

where $\sigma_{\dot{\eta}}$ is the standard deviation of the moonpool velocity.

Following the definition of standard deviation applied to a mass–spring system we can write

$$\sigma_\eta^2 = \int_0^\infty \frac{S_F(\omega)\,d\omega}{(c - m\omega^2)^2 + K^2\omega^2} \tag{3.90}$$

What do S_F, m, K and c mean in this case?

3.7 Steering capability of a ship in following sea

Consider a ship in regular following waves in a situation where the frequency of encounter is zero. Neglect the effect of oscillatory ship motions and diffraction effects of the incident waves. Discuss the steering capability of the ship by discussing the incident velocity to the rudder as a function of the position of the waves relative to the ship.

3.8 Frequency of encounter spectrum

Consider incident long-crested waves on a ship with forward speed U. The wave spectrum is given as $S(\omega_0)$. Assume the frequency of encounter spectrum $S_3^e(\omega_e)$ of heave has been determined.

(a) Explain why the frequency of encounter spectrum $S^e(\omega_e)$ of the incident waves can be related to the wave spectrum by

$$S^e(\omega_e)\,|d\omega_e| = S(\omega_0)\,|d\omega_0|$$

when there is a one-to-one relationship between ω_e and ω_0.

(b) Assume head sea waves and show that

$$S^e(\omega_e) = \frac{S(\omega_0)}{1 + 2U\omega_0/g}$$

How can the transfer function for heave be determined?

(c) Consider a following sea and show that

$$S^e(\omega_e) = \frac{S(\omega_1) + S(\omega_2)}{(1 - 4U\omega_e/g)^{\frac{1}{2}}} + \frac{S(\omega_3)}{(1 + 4U\omega_e/g)^{\frac{1}{2}}}$$

when $\omega_e U/g < \frac{1}{4}$. It is assumed that $\omega_e = |\omega_0 - \omega_0^2 U/g|$.
Illustrate by a figure what the meaning of ω_1, ω_2 and ω_3 is.
What is the result when $\omega_e U/g > \frac{1}{4}$?

4 NUMERICAL METHODS FOR LINEAR WAVE-INDUCED MOTIONS AND LOADS

There exist practical numerical tools based on three-dimensional analyses that predict linear wave-induced motions and loads on large-volume structures at zero Froude number. A wave spectrum is used to describe a sea state and results in an irregular sea can be obtained by linear superposition of results from regular incident waves. Panel methods are the most common techniques used to analyse the linear steady state response of large-volume structures in regular waves. An example of panelling of a TLP is shown in Fig. 4.1, where a total of 12 608 panels is used for the whole structure. In general about 1000 elements would be sufficient. There exist different panel methods. One way is to distribute sources (and sinks) over the mean wetted body surface. Another way is to use a mixed distribution of both sources, sinks and normal dipoles distributed over the mean wetted body surface. Panel methods are also called boundary element methods.

Panel methods are based on potential theory. It is assumed that oscillation amplitudes of the fluid and the body are small relative to cross-sectional dimensions of the body. The effect of flow separation is neglected. This means that the method should not be applied to jacket type structures, risers or tethers. The method can only predict damping due to radiation of surface waves. This means a panel method does not satisfactorily predict rolling motion of a ship close to the roll resonance period because the wave radiation damping moment due to roll may be small and viscous damping effects due to flow separation are important. Another case where panel methods fail to give physically correct answers and viscous effects are important is in predicting vertical forces on a TLP in extreme wave situations in a frequency region where small excitation loads occur. This happens in general in a period range between 15 and 20 s. and is important in establishing design loads for the tethers of a TLP. The reason for the small forces was discussed in connection with equation (3.41). The vertical hydrodynamic forces at the intersection between the pontoons and the columns counteract the hydrodynamic forces on the pontoons. Since the hydrodynamic forces due to potential flow effects become small other physical effects are important. In this case viscous effects matter.

SOURCE TECHNIQUE

We will discuss the source technique in detail and start by repeating what a source is. A source is a point from which fluid is imagined to flow out uniformly in all directions. If the total flux outwards across a small closed surface surrounding the point is Q, then Q is called the 'strength' of the source. A negative source is called a 'sink'. We will from now on use the word source both for sources and sinks.

The velocity potential at any point P, due to a three-dimensional point source, in a liquid at rest at infinity, is

$$\phi = -Q/(4\pi R) \tag{4.1}$$

where R denotes the radial distance of P from the source point. This gives a radial flow from the point and if ds is an element of a spherical surface having its centre at the source, we have the velocity flux through the spherical surface as

$$\iint \frac{\partial \phi}{\partial R} \, \mathrm{d}s = \frac{1}{4\pi} \frac{Q}{R^2} 4\pi R^2 = Q \tag{4.2}$$

In two dimensions the expression for the velocity potential due to a

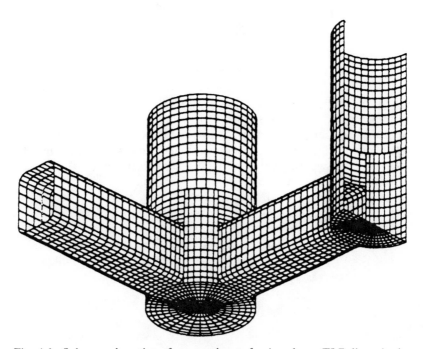

Fig. 4.1. Submerged portion of one quadrant of a six-column TLP discretized with 3152 panels per quadrant (12 608 panels for the total structure) (Korsmeyer *et al.*, 1988.)

point source is

$$\phi = \frac{Q}{2\pi} \log r \tag{4.3}$$

where r is the radial distance from the source point and 'log' means the natural logarithm.

We note that the source expressions are infinite at the source points. However, if we use a continuous representation of sources over a surface the velocity will be finite everywhere in the fluid, except at sharp corners on the body. We will first explain the source technique for a simple problem. We consider a circular cylinder in infinite fluid and we want to find the heave added mass. The boundary value problem we have to solve is partly formulated in Fig. 3.3. The differences are that there is no free-surface and the body boundary condition (3.12) applies for all θ, i.e.

$$\frac{\partial \phi}{\partial r} = -\cos \theta \, |\eta_3| \, \omega \cos \omega t \quad \text{for} \quad r = R \text{ and } -\pi \leqslant \theta \leqslant \pi \tag{4.4}$$

We will find the velocity potential by distributing sources over the body surface. That means we write

$$\phi(y, z) = \int_S q(s) \log((y - \eta(s))^2 + (z - \zeta(s))^2)^{\frac{1}{2}} \, ds \tag{4.5}$$

Here $(\eta(s), \zeta(s))$ are coordinates on the body surface, s is an integration variable along the body surface and (y, z) are coordinates in the fluid domain. The integration is over the wetted body surface S and $q(s)$ is a source density. (Strictly speaking we should have divided the right hand side by 2π if we call $q(s)$ a source density.) Equation (4.5) satisfies the Laplace equation. The source density $q(s)$ is found by satisfying the body boundary condition (4.4). This can be done numerically by the following

1. *Approximate the body surface into N straight lines*
   This is shown in Fig. 4.2 where we have used 16 segments
2. *Assume the source density is constant over each segment*
   This means we approximate equation (4.5) by a sum over each element, i.e.

$$\phi = q_1 \int_{S_1} \log((y - \eta(s))^2 + (z - \zeta(s))^2)^{\frac{1}{2}} \, ds$$

$$+ \ldots$$

$$+ q_{16} \int_{S_{16}} \log((y - \eta(s))^2 + (z - \zeta(s))^2)^{\frac{1}{2}} \, ds \tag{4.6}$$

3. *Satisfy the body boundary condition (4.4) on the mid position $(\bar{y}_i, \bar{z}_i)$ of each segment*

This is done by first normalizing the source density. The purpose of the normalization is to separate out the time-dependence and the unknown heave motion. We write

$$q(s) = -\bar{q}(s) |\eta_3| \, \omega \cos \omega t \qquad (4.7)$$

which expresses the fact that the source density is either in-phase or 180° out-of-phase with the heave velocity. The following linear equation system can be set up for the unknowns $\bar{q}_i$ for each element

$$\bar{q}_1 \left[\frac{\partial}{\partial n} \int_{S_1} [\ ] \, ds \right]_{\bar{y}_1, \bar{z}_1} + \ldots + \bar{q}_{16} \left[\frac{\partial}{\partial n} \int_{S_{16}} [\ ] \, ds \right]_{\bar{y}_1, \bar{z}_1}$$
$$= \cos \theta \, |_{\bar{y}_1, \bar{z}_1}$$

$$\vdots \qquad\qquad \vdots \qquad\qquad\qquad (4.8)$$

$$\bar{q}_1 \left[\frac{\partial}{\partial n} \int_{S_1} [\ ] \, ds \right]_{\bar{y}_{16}, \bar{z}_{16}} + \ldots + \bar{q}_{16} \left[\frac{\partial}{\partial n} \int_{S_{16}} [\ ] \, ds \right]_{\bar{y}_{16}, \bar{z}_{16}}$$
$$= \cos \theta \, |_{\bar{y}_{16}, \bar{z}_{16}}$$

The brackets in the integrals are $\log((y - \eta(s))^2 + (z - \zeta(s))^2)^{\frac{1}{2}}$ and the normal derivative $\partial / \partial n$ is the same as $\partial / \partial r$. When we

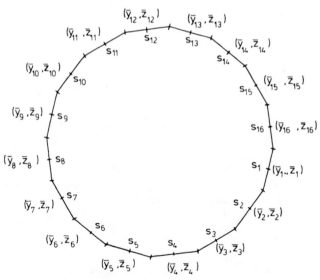

Fig. 4.2. Approximation of a circular cross-section by straight line segments to be used in numerical solutions with the source technique. $((\bar{y}_i, \bar{z}_i) =$ midpoints of segment S_i).

differentiate with respect to n, we should realize that it is y and z that are variables. It means that $n = n(y, z)$ or that $\partial/\partial n \equiv n_2\, \partial/\partial y + n_3\, \partial/\partial z$. Here n_2 and n_3 are the y- and z-components of the normal vector $\mathbf{n}$ to the body surface. In the case of a circular cylinder $n_2 = \sin\theta$, $n_3 = -\cos\theta$ (see Fig. 3.3). Equation (4.8) means we have to solve an equation system of the form

$$A_{ij}\bar{q}_j = B_i \tag{4.9}$$

where both i and j run from 1 to 16. The coefficient matrix in equation (4.9) consists of what we can call influence components. That means if element j has source density 1, it will induce a normal velocity equal to A_{ij} in the midpoint of element i. The total normal velocity at this point is the sum of contributions from each element, and can be expressed by $\sum_{j=1}^{16} A_{ij}\bar{q}_j$. In order to satisfy the boundary conditions this sum shall be equal to the prescribed normal velocity B_i at element i.

4. A normalized velocity potential $\bar{\phi}$ defined by

$$\phi = -\bar{\phi}\,|\eta_3|\,\omega\cos\omega t \tag{4.10}$$

can now be determined from (4.6), (4.7) and the solution $\bar{q}_j$, $j = 1, 16$ found from (4.8).

5. The pressure part that determines the added mass can be found from calculating

$$p = -\rho\frac{\partial\phi}{\partial t} = -\rho\bar{\phi}\,|\eta_3|\,\omega^2\sin\omega t \tag{4.11}$$

6. The resulting vertical force can be found from

$$F_3 = -\int_S pn_3\,\mathrm{d}s \approx -\left\{\rho\sum_{j=1}^{16}\left[\int_{S_j}\bar{\phi}\cos\theta\,\mathrm{d}s\right]\right\}\omega^2\,|\eta_3|\sin\omega t \tag{4.12}$$

By using the definition of added mass and damping (see equation (3.11)) it follows that the two-dimensional added mass in heave can be written as

$$A_{33}^{(2D)} = -\rho\sum_{j=1}^{16}\left[\int_{S_j}\bar{\phi}\cos\theta\,\mathrm{d}s\right] \tag{4.13}$$

and $B_{33}^{(2D)} = 0$ in the infinite domain case.

The integrals in (4.6) and (4.8) can be analytically determined. We will show how this can be done by studying the influence from a source distribution along a segment of the y-axis between 0 and 1. That means that we study

$$\phi(y, z) = \int_0^1 \log((y - \eta)^2 + z^2)^{\frac{1}{2}} \, d\eta \qquad (4.14)$$

The corresponding velocity can be written

$$\frac{\partial \phi}{\partial y} = \int_0^1 \frac{(y - \eta) \, d\eta}{(y - \eta)^2 + z^2} = \frac{1}{2} \log \frac{y^2 + z^2}{(y - 1)^2 + z^2} \qquad (4.15)$$

$$\frac{\partial \phi}{\partial z} = \int_0^1 \frac{z \, d\eta}{(y - \eta)^2 + z^2} = -\frac{z}{|z|} \left[\text{arctg} \frac{y - 1}{|z|} - \text{arctg} \frac{y}{|z|} \right] \qquad (4.16)$$

The vertical velocity $\partial \phi / \partial z$ is plotted in Fig. 4.3 along the line $y = \frac{1}{2}$. Far away from the segment we see that the behaviour is similar to a source singularity in $(0, \frac{1}{2})$. However, close to the segment the behaviour differs, and we see that $\partial \phi / \partial z \to \pi$ when z is approaching a point on the segment from positive values. We translate this information to equation

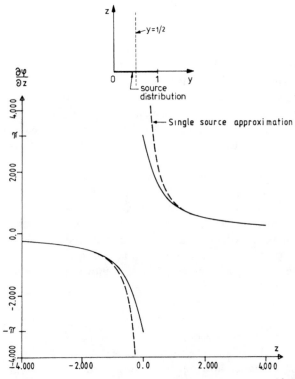

Fig. 4.3. z-variation of the vertical velocity $\partial \phi / \partial z$ induced at $y = \frac{1}{2}$ due to a line distribution of two-dimensional sources of constant strength 2π distributed between $y = 0$ and $y = 1$ on the y-axis.

(4.8) by considering the coordinate system used in equation (4.14) as a local coordinate system for each element. This means that the normal velocity on an element induced by the element itself is equal to π. Another way of saying this is that the diagonal terms A_{ii} in equation system (4.9) are all equal to π.

From equation (4.15) we note that $\partial\phi/\partial y \to \infty$ at the end points $(0, 0)$ and $(0, 1)$ of the segment. This is not a physical phenomenon. It is a consequence of approximating the body by straight line elements and assuming that the source density is constant over each element. However, it illustrates that care should be shown when evaluating fluid behaviour in the vicinity of the body.

Symmetry and antisymmetry properties

It is possible to reduce the equation system (4.8) by considering symmetry properties of the source density. The source density is symmetric about the z-axis because the body is symmetric about the z-axis and is forced to oscillate in heave. This can be shown by noting that the flow is symmetric about the z-axis, i.e. that the y-components of the velocities induced by two sources symmetric about the z-axis have to cancel each other along the z-axis.

Equation system (4.8) can be further reduced by using the fact that the source density is antisymmetric about the y-axis. The reason is that the body is symmetric about the y-axis and is forced to oscillate in heave. Along the y-axis there is only a vertical velocity component. This means the velocity potential is a constant along the y-axis. By setting this constant equal to zero, we see from (4.8) that the source density has to be antisymmetric about the y-axis in order for the velocity potential to be zero along the y-axis.

By using symmetry and antisymmetry properties we have been able to reduce equation system (4.8) from a 16×16 system to a 4×4 equation system. Even if we only satisfy the boundary condition for four points, we must remember to calculate the influence of all 16 segments on each of the four points when we set up the equation system.

In the problem studied above it is not necessary to use a source technique. The velocity potential is well known and can be written as

$$\phi = |\eta_3|\, \omega \cos \omega t \, \frac{R^2}{r} \cos \theta \qquad (4.17)$$

However, analytical solutions cannot be found for general body shapes and numerical methods are then necessary.

One may wonder how equations (4.5) and (4.17) could represent the same solution. For instance, when $r \to \infty$, $\phi \to 0$ according to equation (4.17). It is not evident that equation (4.5) has a similar behaviour. However, this can be shown by using the fact that the source density is antisymmetric about the y-axis and that the distance away from the body is large compared with the cylinder radius. Let us show this by considering two source elements on a body that is symmetric about the y-axis. We can write their contribution to the velocity potential as

$$q(s)\,ds[\log((y - \eta(s))^2 + (z - \zeta(s))^2)^{\frac{1}{2}}$$
$$- \log((y - \eta(s))^2 + (z + \zeta(s))^2)^{\frac{1}{2}}] \quad (4.18)$$

The term in brackets can be expanded in a Taylor expansion about $\eta = 0$, $\zeta = 0$ by considering the field point (y, z) to be fixed and η and ζ to be variables. We find the following expression

$$[\log(y^2 + z^2)^{\frac{1}{2}} - \log(y^2 + z^2)^{\frac{1}{2}}]$$

$$+ \eta\left[\frac{-y}{y^2 + z^2} - \frac{-y}{y^2 + z^2}\right] + \zeta\left[\frac{-z}{y^2 + z^2} - \frac{z}{y^2 + z^2}\right]$$

$$+ \ldots = -2z\zeta/(y^2 + z^2) + \ldots$$

Since $r = (y^2 + z^2)^{\frac{1}{2}}$ is assumed large and ζ is small relative to r, the rest of the terms in the Taylor expansion are negligible relative to $-2z\zeta/(y^2 + z^2)$. By using $z = -r\cos\theta$, we find that (4.18) can be written

$$2\zeta(s)q(s)\,ds\frac{\cos\theta}{r}$$

This is of the same form as (4.17).

Three-dimensional source technique with wave effects

We will show how source technique can be used to analyse linear wave-induced motions and loads on large-volume structures. A ship will be used as an example, and we will start out showing how to find the added mass and damping in heave. It is assumed that the ship has zero forward speed.

The velocity potential ϕ is determined from the following equations:

$$\frac{\partial^2\phi}{\partial x^2} + \frac{\partial^2\phi}{\partial y^2} + \frac{\partial^2\phi}{\partial z^2} = 0 \quad \begin{array}{l}\text{in the fluid domain}\\ \text{on } z = 0\end{array} \quad (4.19)$$

$$-\omega^2\phi + g\frac{\partial\phi}{\partial z} = 0 \quad \begin{array}{l}\text{outside the mean position}\\ \text{of the ship surface}\end{array} \quad (4.20)$$

$$\frac{\partial\phi}{\partial n} = n_3\frac{d\eta_3}{dt} \quad \begin{array}{l}\text{on the mean position}\\ \text{of the ship surface}\end{array} \quad (4.21)$$

$$\frac{\partial\phi}{\partial z} = 0 \quad \text{on} \quad z = -h \qquad \text{for finite water depth}$$
$$\qquad\qquad\qquad\qquad\qquad\qquad\qquad\qquad\qquad\qquad (4.22)$$
$$|\nabla\phi| \to 0 \quad \text{when} \quad z \to -\infty \quad \text{for infinite water depth}$$

Radiation condition   (4.23)

Equation (4.19) is the three-dimensional Laplace equation for the velocity potential. Physically it means that the flow is incompressible and irrotational. Equation (4.20) is the classical linear free-surface condition for steady-state harmonic oscillatory motion of circular frequency ω (see equation (2.15)). Equation (4.21) is the body boundary condition which ensures no flow through the body surface. The unit normal vector $\mathbf{n} = (n_1, n_2, n_3)$ to the ship surface is defined to be positive when pointing into the fluid domain. The bottom condition (equation (4.22)) says that there is no flow through the sea bed. The radiation condition (equation (4.23)) has not been written in mathematical terms, but physically it is necessary to ensure that the waves propagate away from the ship. It may not be obvious a priori why we need a radiation condition. However, in deriving the details of the mathematical solution we will find an ambiguity if we do not enforce the radiation condition. Mathematically there is a possibility both for outgoing and incident waves, but the latter is impossible from a physical point of view. An example of an outgoing wave system in three dimensions and deep water is

$$\phi \sim \frac{Ae^{kz}}{\sqrt{r}} \sin(kr - \omega t + \epsilon) \tag{4.24}$$

Here $r = (y^2 + z^2)^{\frac{1}{2}}$ is a large horizontal distance from the body. If we had written $\sin(kr + \omega t + \epsilon)$ instead of $\sin(kr - \omega t + \epsilon)$ we would have had a circular incoming wave system. We note that ϕ and therefore also the wave amplitude decays like $1/\sqrt{r}$ far away.

It is possible to show that the solution to the boundary-value problem can be represented by a distribution of sources over the mean wetted hull surface. However, the source potential is not the same as we used in infinite fluid, i.e. equation (4.1). The potential has to be corrected so that (4.20), (4.22) and (4.23) are satisfied. Obviously the correction has to satisfy the Laplace equation. The strength of the source density is found by satisfying the body boundary condition (4.21).

For infinite water depth Havelock (1942, 1955) has shown that the source potential can be written as the real part of

$$G(x, y, z; \xi, \eta, \zeta)e^{-i\omega t} = \left[\frac{1}{R} + \frac{1}{R'} \right.$$

$$- \frac{4v}{\pi} \int_0^\infty [v \cos k(z + \zeta) - k \sin k(z + \zeta)] \frac{K_0(kr)}{k^2 + v^2} \, dk$$

$$\left. - 2\pi v e^{v(z+\zeta)} Y_0(vr) + i2\pi v e^{v(z+\zeta)} \mathcal{J}_0(vr) \right] e^{-i\omega t} \tag{4.25}$$

where i is the complex unit and

$$R = ((x - \xi)^2 + (y - \eta)^2 + (z - \zeta)^2)^{\frac{1}{2}}$$
$$R' = ((x - \xi)^2 + (y - \eta)^2 + (z + \zeta)^2)^{\frac{1}{2}}$$
$$r = ((x - \xi)^2 + (y - \eta)^2)^{\frac{1}{2}}$$
$$v = \frac{\omega^2}{g}$$

We should note that the expression is the same if we simultaneously interchange x with ξ, y with η and z with ζ.

Further, $\mathcal{J}_0$ is the Bessel function of the first kind of zero order; Y_0 is the Bessel function of the second kind of zero order; K_0 is the modified Bessel function of zero order, (see Abramowitz & Stegun, 1964). Other ways of writing the source potential can be found in Wehausen & Laitone (1960: pp. 475–9), together with expressions valid for finite water depth. Efficient ways of calculating these sources have been derived by Newman (1985). We can show that equation (4.25) satisfies the radiation condition in the form (4.24) by using asymptotic expansions for Bessel functions for large r-values (see Abramowitz & Stegun, 1964). We can write

$$\mathrm{Re}\{(-Y_0(vr) + i\mathcal{J}_0(vr))e^{-i\omega t}\}$$

$$\approx \mathrm{Re}\left\{\left[-\left(\frac{2}{\pi vr}\right)^{\frac{1}{2}} \sin\left(vr - \frac{\pi}{4}\right)\right.\right.$$
$$\left.\left. + i\left(\frac{2}{\pi vr}\right)^{\frac{1}{2}} \cos\left(vr - \frac{\pi}{4}\right)\right]e^{-i\omega t}\right\}$$
$$= -\left(\frac{2}{\pi vr}\right)^{\frac{1}{2}} \sin\left(vr - \omega t - \frac{\pi}{4}\right)$$

where Re denotes the real part. This term is order of magnitude $r^{-\frac{1}{2}}$ (i.e. $O(r^{-\frac{1}{2}})$) for large values of r. We easily see that the two first terms in the brackets in equation (4.25) are $O(r^{-1})$ for large values of r. Since K_0 is exponentially small for large r and the integrand is always finite, the third term in the brackets can be shown to be negligible compared with the last two terms. This means we have shown that the source potential represents outgoing waves far away from the source point. It is also possible to show that the free-surface condition (equation (4.20)) is satisfied. Fig. 4.4 shows a picture of the wave system caused by a source.

It is possible to show from equation (4.25) that

$$Ge^{-i\omega t} \to \left[\frac{1}{R} + \frac{1}{R'}\right]e^{-i\omega t} \quad \text{when} \quad \omega \to 0 \qquad (4.26)$$

and

$$Ge^{-i\omega t} \rightarrow \left[\frac{1}{R} - \frac{1}{R'}\right]e^{-i\omega t} \quad \text{when} \quad \omega \rightarrow \infty \qquad (4.27)$$

The latter means that the integral in (4.25) approaches $-2/R'$ when $\omega \rightarrow \infty$.

Equation (4.26) expresses the source potential as a combination of a source in infinite fluid and an image source above the free-surface when $\omega \rightarrow 0$. (To be precise $1/R$ is actually a sink with density 4π (see equation (4.1)). However, we use the word 'source' for $1/R$ in this context.) This confirms that the velocity potential satisfies the rigid wall condition $\partial\phi/\partial z = 0$ on $z = 0$ when the frequency of oscillation goes to zero. This is illustrated in Fig. 4.5.

Equation (4.27) expresses the fact that the source potential is a combination of a source in infinite fluid and an image sink above the free-surface when $\omega \rightarrow \infty$. This expression satisfies the free-surface condition $\phi = 0$ on $z = 0$. This is illustrated in Fig. 4.6. The solution of the velocity potential for the forced heave problem can be written as a distribution of sources over the mean wetted hull surface S_B as

$$\phi(x, y, z; t) = \text{Re}\left\{\iint_{S_B} dsQ(s)G(x, y, z; \xi(s), \eta(s), \zeta(s))e^{-i\omega t}\right\} \qquad (4.28)$$

where the source function $Ge^{-i\omega t}$ is given by equation (4.25) for the

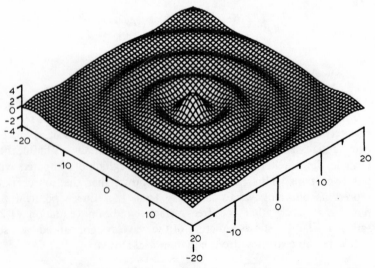

Fig. 4.4. Picture of the wave field caused by a harmonically oscillating source with zero mean speed.

infinite water depth problem. $Ge^{-i\omega t}$ is also called Green function. The source density $Q(s)$ is complex with a real and an imaginary part and is found by satisfying the body boundary condition. This is done by first introducing a normalized source density $\bar{Q}(s)$ so that

$$Q(s)e^{-i\omega t} = \bar{Q}(s)\omega \,|\eta_3|\, e^{-i\omega t} \qquad (4.29)$$

The purpose of the normalization is, as before, to separate out the time dependence and the unknown heave motion. The body boundary condition leads to an integral equation for $\bar{Q}(s)$ which in general we are not able to solve analytically. What we do is to discretize the problem by dividing the hull surface into elements, over which we assume the source density is constant. Fig. 4.7 shows an example of how a ship hull surface

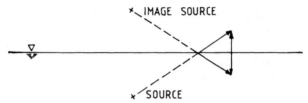

Fig. 4.5. Illustration that a wave source satisfies the rigid wall condition on the free-surface when the frequency of oscillation $\omega \to 0$. The wave source consists of a source in infinite fluid and an image source. The field point is in the mean free-surface. The solid line that is in the same direction as the dashed line from the source point shows the velocity vector due to the source. The velocity vector due to the image source is illustrated in a similar way. The vertical solid lines are the vertical components of the velocity vectors.

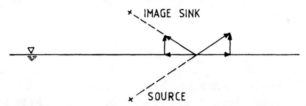

Fig. 4.6. Illustration that a wave source satisfies the zero potential condition on the free-surface when the frequency of oscillation $\omega \to \infty$. The wave source consists of a source in infinite fluid and an image sink. The field point is in the mean free-surface. The solid line that is in the same direction as the dashed line from the source point shows the velocity vector due to the source. The velocity vector due to the image sink is illustrated in a similar way. The horizontal and vertical solid lines are respectively the horizontal and vertical components of the velocity vectors. The total horizontal velocity is zero on the free-surface, which means the velocity potential is a constant on the free-surface. The constant is chosen to be zero.

can be approximated by plane quadrilateral elements. The number of elements N is 810 in this case. The discretization of the problem leads to a linear equation system for the unknown source density in a similar way to the forced heave problem for a circular cylinder in infinite fluid. The differences, which we will point out, are

1. The source density is complex in the wave problem.
   Physically it means that the source density is generally not in-phase or 180° out-of-phase with the heave velocity as it is in an infinite fluid problem (see equation (4.7)). This means there are N real and N imaginary unknowns or N complex unknowns if the body is approximated by N elements and no symmetry and antisymmetry properties are accounted for. Symmetry and antisymmetry properties can be accounted for in a similar way to the two-dimensional problem. This means the equation systems written in complex variables are reduced to $N/2 \times N/2$ or $N/4 \times N/4$ with respectively one or two symmetry planes.

2. The source expression is far more complicated to compute numerically in a wave problem compared with an infinite fluid problem (see equation (4.25) compared to equation (4.1)).

3. The normal velocity induced on an element by a source distribution over the same element is different in a two-dimensional and a three-dimensional problem (see exercise 4.1). We find

$$\left[\frac{\partial}{\partial n} \int_{S_i} [(x - \xi(s))^2 + (y - \eta(s))^2 + (z - \zeta(s))^2]^{-\frac{1}{2}} \, ds \right]_{\bar{x}_i, \bar{y}_i, \bar{z}_i} = -2\pi \quad (4.30)$$

Here $(\bar{x}_i, \bar{y}_i, \bar{z}_i)$ is the geometrical area centre of element S_i.

4. A solution may not exist for all frequencies if the potential is represented as a distribution of wave sources over the body surface. For a surface piercing body there exist an infinite number of discrete frequencies (*irregular frequencies*) that cause the three-dimensional source technique to break down (John, 1950). The assumption in John's analysis is that no point of the immersed surface would be outside a cylinder

Fig. 4.7. Approximation of the submerged part of a ship hull by plane quadrilateral elements. Total number of elements is 810.

drawn vertically downward from the intersection of the body with the free-surface, and that the free-surface would be intersected orthogonally by the body in its mean or rest position. A ship without a bulbous bow will for instance normally satisfy this assumption. However, it is possible that John's analysis applies to a broader class of bodies than he restricted himself to. John shows that the lowest irregular circular frequency ω_1 satisfies $\omega_1(D/g)^{\frac{1}{2}} \geqslant 1.0$, where D is the draught of the structure. If we translate this to a ship at zero forward speed it means that we are sure there are no irregular frequencies when $\lambda/L > 2\pi D/L$ where L is the ship length. In practice the irregular frequencies do not represent a serious problem for analysis of ship motions at zero forward speed. The wavelengths of importance in predicting ship motions are not in a range where irregular frequencies occur. For a vertical cylinder standing on the sea bed and penetrating the free-surface, the wavelength corresponding to the lowest irregular frequency for surge and heave is respectively $0.82D$ and $1.31D$ with D being the diameter. In practice, the solution will also be unsatisfactory in the neighbourhood of the irregular frequencies.

We must emphasize that the irregular frequencies are not caused by any physical phenomenon. Therefore if a different technique is used, a solution may be found at the irregular frequencies for the source technique. An example of a different technique is a purely analytical solution. This exists in very special cases, for instance to calculate the wave loads on a fixed vertical cylinder standing on the sea bed and penetrating the free-surface (McCamy & Fuchs, 1954).

Irregular frequencies in the source technique represent eigenfrequencies for a *fictitious* fluid motion inside the body with the same free-surface condition as outside the body and the body boundary condition $\phi = 0$. In general it is difficult to find the eigenfrequencies analytically before one solves the exterior physical problem. The word *fictitious* for the interior flow is stressed. The flow is not a physical 'sloshing' motion. In the sloshing case the body boundary condition would be $\partial\phi/\partial n = 0$. However, to get an idea what the irregular frequencies might be we should have the sloshing modes in mind. For a submerged body no irregular frequencies exist.

What is happening mathematically at an irregular frequency is that the determinant of the coefficient matrix in the linear equation system for the unknown source densities goes to zero when the number of unknowns goes to infinity.

Lee & Sclavounos (1989) have presented a practical way to avoid the problem of irregular frequencies. If no attempts are made to get rid of the irregular frequencies, they may be difficult to detect if few frequencies are used in the calculations. Even if many frequencies are used, the irregular frequencies may be difficult to detect in special cases. An example of this is calculation of added mass and damping for a catamaran. In this case the strong physical interaction effects between the hulls can be mixed up with irregular frequency behaviour.

5. *Panelling of the hull surface*

The panelling of the hull surface is often done by plane quadrilateral elements in three-dimensional problems. Details of how to create plane quadrilateral elements mathematically have been presented by Hess & Smith (1962). Dependent on the body shape the panels may create 'leaks' through the body surface. The reason is that the panels do not fit together. This may disturb a practical maritime person, but has no serious consequences for the physical description of the fluid. The condition of no flow through the hull surface is satisfied at the geometrical area centre of each panel and nothing is required about the fluid behaviour at the edges of elements where geometrical 'leaks' between the elements may appear.

There is no unique way to approximate the hull surface by elements. However, one should keep in mind that one assumes the source density and the fluid pressure to be constant over each element. Therefore one should keep smaller elements in areas where the flow changes more rapidly. An example of the latter is in the vicinity of sharp corners. It should be realized that the numerical solution for velocities never is satisfactory on the element closest to a sharp corner. The reason is that the potential flow solution has a singularity there, and that this is inconsistent with assuming the source density and velocity potential to be constant over an element. In reality the flow will separate at a sharp corner. This effect is not included in the method. In the wave zone the element size should be small compared to the wavelength. A characteristic length of an element ought to be at most $\frac{1}{8}$ of the wavelength. Around a vertical column with a circular cross-section there ought to be 15–20 circumferential elements at any height. If there is a conflict between these two recommendations, the more conservative is required.

The body boundary condition is often satisfied on the geometrical mid-point. The elements must not be selected so that any element mid-points are very close to the edges of another element. The reason is that induced velocities from an element are singular at its edges (see similar discussion of induced velocity from a source element in two-dimensional flow). Problems like this may happen for very thin bodies or for small gap problems.

Typical values for the total number of elements may vary from 500 to 1500. However, Fig. 4.1 shows an example where 12 608 panels were used. This requires iterative solutions of the linear equation systems for the source densities in order to be within practical limits of CPU time on available computer hardware. The best way to find a sufficient number of elements, is to do calculations with increasing numbers of elements and check convergence of the results.

The solution procedure that we have set up can be generalized to any modes of motion, i.e. to surge, sway, roll, pitch and yaw motion (see for instance Faltinsen & Michelsen, 1974). This means we can find the added mass and damping matrix. We may also solve the diffraction problem by using the body boundary condition $\partial\phi/\partial n = -(\partial\phi^I/\partial n)$ where ϕ^I is incident wave potential. This enables us to find the wave excitation loads, i.e. wave excitation forces in surge, sway and heave as well as wave excitation moments in roll, pitch and yaw. We can then set up the equation system for solving motions in six degrees of freedom (see equation (3.47)). When the motions are found, we may calculate flow details by using the expression for the velocity potential in terms of a three-dimensional source distribution. The free-surface elevation can be evaluated by using the free-surface condition

$$g\zeta + \left.\frac{\partial\phi}{\partial t}\right|_{z=0} = 0 \qquad\qquad (4.31)$$

By Bernoulli's equation it follows that the dynamic pressure is $p = -\rho\,\partial\phi/\partial t$ at the mean position of the body. When we calculate the dynamic pressure at the hull for one particular time instant, we should include the hydrostatic pressure $-\rho gz$ and take into account that z is changed due to the vertical rigid body motion of the hull. Note, however, that it is consistent with linear theory to evaluate $\partial\phi/\partial t$ at the mean position in the calculation of the pressure. Care should be shown in evaluating velocities close to the body. One reason is that the velocities are singular at the edges of the element. Experience has shown that one has to be of the order of a characteristic element length outside the body in order to always obtain satisfactory estimations of fluid velocities. Further, the velocity potential should not be evaluated at places on the element other than at the geometrical area centre.

Source methods have been used in practical calculations of wave loads on large volume offshore structures for about twenty years. However, recent comparative studies performed by ISSC and ITTC have shown large variations in results by different computer programs with the same theoretical foundation. Causes of differences can be grid shape, size and distribution, geometry approximation, singularity density distribution over each panel, Green function calculation, and how singularities are integrated over panels. However, a well documented source method represents a valuable tool in engineering analysis of offshore structures.

ALTERNATIVE SOLUTION PROCEDURES

It is not necessary to use plane quadrilateral elements and assume constant source density over each element. A different procedure is to use triangular elements and assume a linear variation over each element (Bai & Yeung, 1974). This has the advantage of obtaining a better approximation of the hull surface. The body boundary conditions are satisfied at the vertices of the elements, which requires special care.

There exist panel methods other than the source technique. One method uses Green's second identity which states

$$\iiint\limits_{\Omega} (\phi \nabla^2 \psi - \psi \nabla^2 \phi) \, d\tau = \iint\limits_{S} \left(\psi \frac{\partial \phi}{\partial n} - \phi \frac{\partial \psi}{\partial n} \right) dS \qquad (4.32)$$

where S is the surface enclosing the fluid volume Ω. It is necessary that ϕ and ψ have continuous derivatives of first and second order in Ω. The normal direction n is into the fluid region.

Example 1

If $\nabla^2 \phi = 0$ and $\nabla^2 \psi = 0$ everywhere in the fluid domain it follows from equation (4.32) that

$$\iint\limits_{S} \left(\psi \frac{\partial \phi}{\partial n} - \phi \frac{\partial \psi}{\partial n} \right) ds = 0 \qquad (4.33)$$

Example 2

If $\nabla^2 \phi = 0$ everywhere in the fluid domain,

$$\psi(x, y, z; x_1, y_1, z_1) = \{(x - x_1)^2 + (y - y_1)^2 + (z - z_1)^2\}^{-\frac{1}{2}} \qquad (4.34)$$

and the point (x_1, y_1, z_1) is inside the fluid volume (see Fig. 4.8), we have to be careful how we handle the singular point (x_1, y_1, z_1) in equation (4.32). We note that ψ is the velocity potential for a source of strength -4π with a source at point (x_1, y_1, z_1). Outside the singular point ψ satisfies the Laplace equation. Equation (4.33) applies therefore if we integrate over $S + S_1$, i.e.

$$\iint\limits_{S+S_1} \left(\psi \frac{\partial \phi}{\partial n} - \phi \frac{\partial \psi}{\partial n} \right) ds = 0 \qquad (4.35)$$

where S_1 is the surface of a sphere with small radius r enclosing the point (x_1, y_1, z_1). At S_1 we can write $\partial \psi / \partial n = \partial \psi / \partial r = -1/r^2$, i.e.

$$\iint\limits_{S_1} \phi \frac{\partial \psi}{\partial n} ds \approx 4\pi r^2 \left(-\frac{1}{r^2} \right) \phi(x_1, y_1, z_1) \qquad (4.36)$$

$$\iint\limits_{S_1} \frac{\partial \phi}{\partial n} \psi \, ds \approx 0 \qquad (4.37)$$

That means we can write the velocity potential in three-dimensional flow as

$$4\pi\phi(x_1, y_1, z_1) = \iint\limits_{S} \left[\phi(s) \frac{\partial}{\partial n} \left(\frac{1}{R} \right) - \frac{1}{R} \frac{\partial \phi(s)}{\partial n} \right] ds \qquad (4.38)$$

Here S has to be a closed surface. If we apply the formula to the wave load problem, the surface S consists of the mean body surface S_B, a vertical circular cylindrical surface S_∞ away from the body, the mean free surface S_F and sea bottom S_0 inside S_∞ (see Fig. 4.9). Equation (4.38) states that the velocity potential can be represented by a distribution of sources R^{-1} (also called Rankine sources) and dipoles $(\partial / \partial n)(1/R)$ over the closed surface $S = S_B + S_F + S_\infty + S_0$. In the differentiation $(\partial / \partial n)(1/R) = \partial / \partial n [(x - x_1)^2 + (y - y_1)^2 + (z - z_1)^2]^{-\frac{1}{2}}$ it is x, y and z that are considered as variables. The dipole density on S is given by $\phi(s)$ and the source density is given by $\partial \phi / \partial n$.

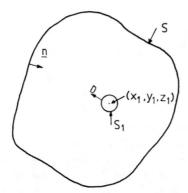

Fig. 4.8. Integration surfaces used in applying Green's second identity. ((x_1, y_1, z_1) = singular point of equation (4.34).)

One has to be careful when (x_1, y_1, z_1) is at the boundary S. The reason is the integration $(\partial/\partial n)(1/R)$ in the vicinity of (x_1, y_1, z_1). One often sees that the factor 4π on the left hand side of equation (4.38) is replaced by 2π when (x_1, y_1, z_1) is on S. This takes care of the contribution of the integration of $(\partial/\partial n)(1/R)$ in the close vicinity of (x_1, y_1, z_1).

Equation (4.38) is a formal representation of the velocity potential. In addition boundary conditions and radiation conditions have to be imposed (see for instance equations (4.20) to (4.23)).

The infinite fluid source $1/R$ and dipole $(\partial/\partial n)(1/R)$ in equation (4.38) can be replaced by a source and a dipole satisfying the free-surface condition (equation (4.20)), the bottom condition (equation (4.22)) and the radiation condition. The source expression in infinite water depth is given by equation (4.25). If this is done it is possible to show that equation (4.38) is reduced to a distribution over the body surface S_B.

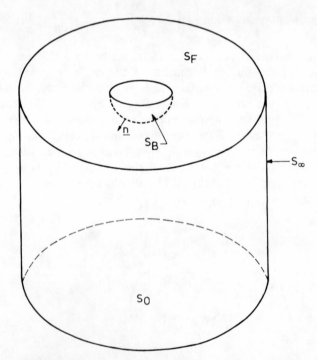

Fig. 4.9. Integration surfaces used in connection with a Green's second identity representation of the velocity potential in terms of a distribution of Rankine (infinite fluid) sources and dipoles. The surfaces are the mean wetted body surface S_B, a vertical cylindrical control surface S_∞ extending from the mean free surface to the sea bottom, the mean free surface S_F and sea bottom S_0 inside S_∞.

We can show this for deep water by defining

$$\psi(x, y, z; x_1, y_1, z_1) = \{[(x - x_1)^2 + (y - y_1)^2 + (z - z_1)^2]^{-\frac{1}{2}}$$
$$+ f(x, y, z; x_1, y_1, z_1)\}e^{-i\omega t}$$
$$\equiv G(x, y, z; x_1, y_1, z_1)e^{-i\omega t}$$

where the source potential is given by equation (4.25). We start out with equation (4.35). Since f is regular everywhere in the fluid, it does not contribute to the integral over S_1 when $r \rightarrow 0$. This means

$$\iint \phi \frac{\partial \psi}{\partial n} ds = -4\pi\phi(x_1, y_1, z_1)e^{-i\omega t}$$

On the mean free-surface S_F we can write

$$\iint_{S_F} \left(\psi \frac{\partial \phi}{\partial n} - \phi \frac{\partial \psi}{\partial n}\right) ds = -\iint_{S_F} \left(\psi \frac{\partial \phi}{\partial z} - \phi \frac{\partial \psi}{\partial z}\right) ds$$
$$= -\frac{\omega^2}{g} \iint_{S_F} (\psi\phi - \phi\psi) ds = 0$$

We have here used the fact that ψ and ϕ satisfy the same free-surface condition.

On S_∞ far away from the body both ϕ and ψ represent outgoing waves as in equation (4.24). We can then show that the integral over S_∞ disappears. At the surface S_0 far down in the fluid there is no disturbance and therefore no contribution from the integration over S_0. The total result is that we can write

$$\phi(x_1, y_1, z_1) = \frac{1}{4\pi} \iint_{S_B} \left(\phi \frac{\partial G}{\partial n} - G \frac{\partial \phi}{\partial n}\right) ds \qquad (4.39)$$

This type of representation of the velocity potential is used by Newman & Sclavounos (1988) where more details about the solution procedure of the unknown velocity potential on the body surface may be found.

The representation of the velocity potential in terms of infinite fluid sources and dipoles has a great advantage when the free-surface condition is so complicated that it is not possible to find an analytical expression for the wave source that satisfies the free-surface condition. An example of this is if the completely non-linear free-surface conditions are used. However, a disadvantage with having a representation like equation (4.38) with unknown velocity potential distribution over S_B, S_F, S_∞ and S_0 is that it will lead to a large equation system for solution of the unknown potential distribution. A better way is to use a hybrid technique, where an analytical expression is used outside S_∞, and S_∞ is kept as close as possible to the body surface. This was used by Zhao et *al.* (1988) in studying wave–current interaction on large volume structures. In this case it was not possible to find an expression for a wave source that satisfied the free-surface condition close to the body.

We have only mentioned boundary element methods above. A different approach is to use a finite element method in combination with the analytical representation outside S_∞ (Mei, 1983). Yeung (1982) has given a review article on numerical methods applicable to wave–body interaction problems.

FORWARD SPEED AND CURRENT EFFECTS

When a ship has forward speed or there is a current present around a structure, the free-surface condition (4.20) does not apply. We will derive the correct form in the following text.

The velocity potential ϕ can be separated into two parts; one is the time-independent steady contribution due to the forward motion of the structure and the other the time-dependent part associated with the incident wave system and the unsteady body motion. We will use a reference frame moving with the forward speed of the structure. It is assumed the incident wave amplitudes are sufficiently small so that linear theory applies for the unsteady wave effects. We can formally write

$$\Phi(x, y, z; t) = [Ux + \phi_s(x, y, z)] + \phi_T(x, y, z)e^{-i\omega_e t} \qquad (4.40)$$

Here $Ux + \phi_s$ is the steady contribution with U being the forward speed of the ship. Physically, $Ux + \phi_s$ describe the steady flow including the steady wave pattern created around a ship. Further, ϕ_T is the complex amplitude of the unsteady potential, and ω_e is the circular frequency of encounter in the moving reference frame. It is understood that the real part is to be taken in expressions involving $e^{-i\omega_e t}$.

The free-surface conditions state that the pressure is equal to atmospheric pressure on the free-surface and that a particle on the free-surface remains on the free-surface (see chapter 2). Bernoulli's equation (see equation (2.4)) is used to calculate the pressure p, i.e.

$$p + \rho \frac{\partial \Phi}{\partial t} + \rho g z + \frac{\rho}{2}\left[\left(\frac{\partial \Phi}{\partial x}\right)^2 + \left(\frac{\partial \Phi}{\partial y}\right)^2 + \left(\frac{\partial \Phi}{\partial z}\right)^2\right] = C \quad (4.41)$$

The constant C in the equation is chosen as the value of the left hand side far away from the body, where there is no free-surface elevation and $\Phi = Ux$, i.e.

$$C = \frac{\rho}{2} U^2 + p_0 \qquad (4.42)$$

By neglecting the interactions with the steady velocity potential ϕ_s we can write the linearized dynamic and kinematic free surface conditions as

$$g\zeta - i\omega_e\phi_T + U\frac{\partial\phi_T}{\partial x} = 0 \quad \text{on} \quad z = 0 \tag{4.43}$$

$$-i\omega_e\zeta + U\frac{\partial\zeta}{\partial x} - \frac{\partial\phi_T}{\partial z} = 0 \quad \text{on} \quad z = 0 \tag{4.44}$$

Here $\zeta e^{-i\omega_e t}$ is the unsteady wave elevation. Equations (4.43) and (4.44) can be combined to

$$\left[\left(-i\omega_e + U\frac{\partial}{\partial x}\right)^2 + g\frac{\partial}{\partial z}\right]\phi_T = 0 \quad \text{on} \quad z = 0 \tag{4.45}$$

A wave source satisfying equation (4.45) and the radiation condition give a wave picture that is strongly influenced by the forward speed. In the zero speed case there is only one wave system generated by the source (see Fig. 4.4). The wave system generated by a source at forward speed is more complicated than in the zero-speed case. There are several different wave systems. Let us show this with two examples for the deep water case. One case is for $\tau = \omega_e U/g < \frac{1}{4}$ and the other case is for $\tau > \frac{1}{4}$ (see Figs 4.10 and 4.11). When $\tau > \frac{1}{4}$ there is no far-field wave effect straight ahead of a translatory harmonic oscillating source. Mathematical expressions for the sources may be found in Wehausen & Laitone (1960: pp. 490–5). Figs. 4.10 and 4.11 give crest numbers associated with the different wave systems AA, D1, D2. There are two crest numbers given and two 'crests' shown for each wave system. What we call 'crests' are of course only real crests for specific time instants with periodic intervals. The difference between the two wave crest numbers for each wave system gives how many 'crests' there are between the two 'crests' which we have shown for each wave system AA, D1, D2. For instance, in Fig. 4.10 there are $(138 - 92) = 46$ 'crests' between the two 'crests' shown for the wave system AA. For each wave system we may have both divergent and transverse waves. The latter notation is commonly used for the two types of waves we see behind a ship in steady forward motion in calm sea. Fig. 4.12 shows an example of numerical calculated diffraction waves at one time instant around a restrained ship in head sea waves when $\tau > \frac{1}{4}$. It should be noted that the incident waves should be added to give the total wave picture around the ship. The reason why the waves are left downstream of a ship when $\tau > \frac{1}{4}$ can be understood in terms of the group velocity or the energy propagation velocity of the different wave systems. If we regard the problem from an earth-fixed coordinate system, all the waves have a smaller group velocity than the ship speed U when $\tau > \frac{1}{4}$ and can therefore not propagate ahead of the ship. At $\tau = \frac{1}{4}$ the group velocities of parts of the unsteady ship-generated waves are precisely equal to U. For $\tau < \frac{1}{4}$ some of the waves can propagate out ahead of the ship. In the two-dimensional flow case this is discussed in more detail in exercise 4.4(b). The two-dimensional case is also discussed by Grue & Palm (1985).

The effect of current on a structure is similar to the effect of forward speed on a ship. However, practical cases often occur for $\tau = \omega_e U/g < \frac{1}{4}$. The τ-value is so small that the free-surface condition (4.45) can be approximated by

$$-\omega_e^2 \phi_T - 2i\omega_e U \frac{\partial \phi_T}{\partial x} + g \frac{\partial \phi_T}{\partial z} = 0 \quad \text{on} \quad z = 0 \qquad (4.46)$$

A consequence of using (4.46) in combination with the radiation condition is that there is only one wave system present. This is illustrated in Fig. 4.13. We note that the wavelength becomes longest when the local wave propagation direction coincides with the current direction and

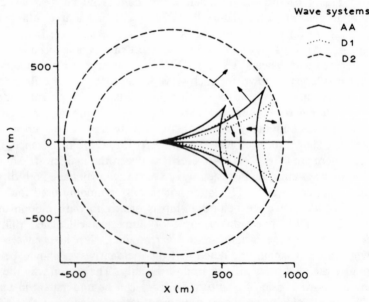

DEEP WATER
Velocity $U = 2{\cdot}50\ \text{m s}^{-1}$ $\omega_e = 0{\cdot}5\ \text{rad s}^{-1}$
$\tau = 0{\cdot}1274$

Wave systems

—— AA
········ D1
––––– D2

Fig. 4.10. Picture of the wave field caused by a harmonically oscillating source with mean forward speed U in deep water when $\tau < \frac{1}{4}$, ($\tau = \omega_e U/g$, ω_e = circular frequency of oscillation). The crest numbers shown for wave system AA are 92 and 138, for wave system D1, 168 and 252 and for wave system D2, 2 and 3. (Børresen, 1984.)

shortest when the local wave propagation direction is opposite to the current direction. It should be noted that the analysis is based on $\tau = \omega_e U/g$ being small and below $\frac{1}{4}$.

Analysis of wave–current interaction on offshore structures is presented by Zhao et al. (1988b). Satisfactory agreement between experimental and numerical values was documented. In the forward motion problem of ships at $\tau > \frac{1}{4}$ there are still many unsolved problems and three-dimensional panel methods with combined forward speed and wave effects do not give any better agreement with experimental values for vertical ship motions than strip theories. To make further improvements in ship motion predictions at moderate and high Froude number it is felt that one first has to study the steady wave potential problem in more detail. This is referred to as the wave resistance problem in ship hydrodynamics and is known to be a difficult numerical and physical problem. It is fair to say that there exists at present no general practical numerical method for the wave resistance problem.

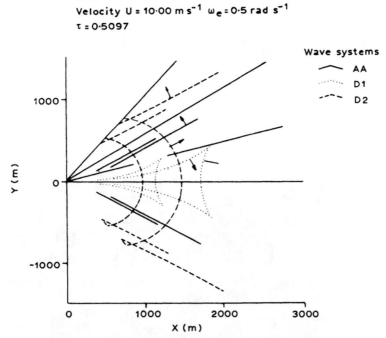

Fig. 4.11. Picture of the wave field caused by a harmonically oscillating source with mean forward speed in deep water when $\tau > \frac{1}{4}$. For notation see Fig. 4.10. The crest numbers shown for wave system AA are 2 and 3, for wave system D1 34 and 51 and for wave system D2 2 and 3. (Børresen, 1984.)

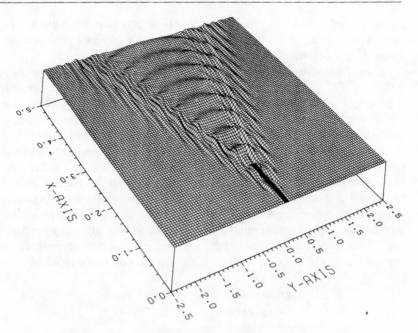

Fig. 4.12. Diffraction waves at one time instant around a restrained ship in head sea waves. Deep water. $F_n = U/\sqrt{(Lg)} = 0.2$, $\lambda/L = 0.5$, $\tau = 1.212$. (L = ship length, λ = incident wavelength, U = forward speed, $\tau = \omega_e U/g$, ω_e = circular frequency of oscillation (encounter frequency).) (Ohkusu & Iwashita, 1987.)

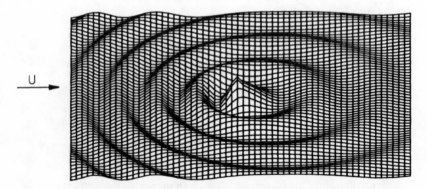

Fig. 4.13. Picture of the wave field caused by a harmonically oscillating source with mean forward speed U satisfying the free-surface condition (4.46) and valid for $\tau < \frac{1}{4}$ and deep water. ($\tau = \omega_e U/g$, ω_e = circular frequency of oscillation.)

EXERCISES

4.1   Induced velocities by a three-dimensional source element
Consider a source distribution over a plane circular element of radius R (see Fig. 4.14). Assume the fluid has infinite extent and the source density is constant and equal to one. Show that the vertical velocity V_z at a point p on the z-axis (see Fig. 4.14) induced by the source distribution can be written as

$$V_z = -0.5 \left[\frac{z}{(R^2 + z^2)^{\frac{1}{2}}} - \frac{z}{|z|} \right]$$

Discuss the velocity behaviour close to the circular element both for negative and positive z-values. What is the behaviour like far away from the source element?

4.2   Motions and loads on a barge
Consider the barge presented in Fig. 3.25. By means of three-dimensional source technique we want to find wave-induced motions and loads when the forward speed is zero and there is no current present.

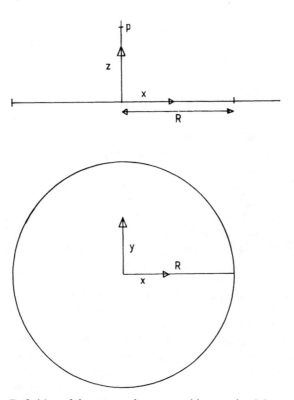

Fig. 4.14. Definition of the source element used in exercise 4.1.

(a) What are the irregular frequencies if the barge has infinite length? (Hint: Solve the interior problem as in exercise 2.1, but with the body boundary condition $\phi = 0$.)

Answer: $\dfrac{\omega_j^2}{g} = j\dfrac{\pi}{B}\coth\left(\dfrac{j\pi D}{B}\right), \quad j = 1, 2, \ldots$

(b) Will irregular frequencies represent a problem?
(c) Discuss and show how to select panels for a three-dimensional source analysis.
(d) How many symmetry planes can we benefit from in the hydrodynamic calculations?

4.3 Green's second identity

(a) Show for two-dimensional flow that the velocity potential ϕ can be written as

$$\phi(x_1, y_1) = -\frac{1}{2\pi}\int_S \left(\phi\frac{\partial}{\partial n}\log r - \frac{\partial\phi}{\partial n}\log r\right) ds \qquad (4.47)$$

where $r = ((y - y_1)^2 + (z - z_1)^2)^{\frac{1}{2}}$, S is a surface enclosing a fluid volume and (y_1, z_1) is a point inside the fluid volume. Further, $(y(s), z(s))$ are points on surface S and $\partial/\partial n \equiv n_2\,\partial/\partial y + n_3\,\partial/\partial z$ where the unit normal vector $\mathbf{n} = (0, n_2, n_3)$ to S is defined to have positive direction into the fluid domain.

(b) Consider a two-dimensional body near a plane wall. Start out with Green's second identity and show how the velocity potential can be written as a distribution of sources and dipoles over the body surface only.

(c) Consider the problem of finding the velocity potential for forced linear heave motion of a two-dimensional surface piercing body when the frequency of oscillation goes to zero. Use the result from question (b) and show that the dominant behaviour far away from the body is source-like. What is the source strength? (Hint: Use the Taylor expansion of $\partial/\partial n(\log r)$ and $\log r$ with y and z as variables.)

4.4 Forward speed effects on wave systems in 2D flow

Consider an infinitely long horizontal cylinder in a steady incident flow and a beam sea with regular waves in deep water. Potential flow theory is assumed. The cylinder is restrained from drifting. Due to interaction between the incident waves and the body, the body will generate waves. The steady incident velocity potential far away from the body is written as in equation (4.40) with $\phi_s = 0$. This means it represents a steady flow with constant velocity along the positive x-axis. The time dependence of the oscillatory flow is expressed as $e^{-i\omega_e t}$. At some distance from the body the time dependent velocity potential satisfies the free-surface condition (equation (4.45)). The body will generate different wave systems far away from the body. The velocity potential of these wave systems has to be of the form

$$\phi_T e^{-i\omega_e t} = A e^{kz \pm ikx - i\omega_e t} \tag{4.48}$$

where A is constant, in order that the two-dimensional Laplace equation should be satisfied. The k-value is real and positive in order for the flow to vanish far down in the fluid and for the velocity potential to represent waves at large positive or negative x-values.

(a) Show that

$$k_{1,2} = \frac{g}{2U^2}(1 - 2\tau \pm (1 - 4\tau)^{\frac{1}{2}})$$

when ϕ is proportional to $e^{-ikx - i\omega_e t}$ and $\tau = \omega_e U/g \leqslant \frac{1}{4}$, and that

$$k_{3,4} = \frac{g}{2U^2}(1 + 2\tau \pm (1 + 4\tau)^{\frac{1}{2}})$$

when ϕ is proportional to $e^{ikx - i\omega_e t}$.

(b) Consider the problem from a relative frame of reference system $x' = x - Ut$. In this coordinate system the body moves with a velocity U in the negative x'-direction. Consider the different wave systems in this coordinate system. The group velocity (energy propagation velocity) of the different wave systems can be written as

$$C_g = \frac{1}{2}\sqrt{\frac{g}{k}}$$

If the group velocity is less than U in the direction of forward motion, it means the wave system has to be on the downstream side of the object. Discuss what wave systems are upstream or downstream of the body. What happens when $\tau = \frac{1}{4}$?

(c) Assume that the wave numbers k_1, k_2, k_3 and k_4 can be used as an approximation for estimating the wavelengths of the wave systems along the track of a three-dimensional harmonically oscillating source with mean forward speed in deep water. Determine how this agrees with the results in Figs. 4.10 and 4.11. Discuss the phase velocity direction of the transverse wave systems and check if this agrees with the arrows in Figs. 4.10 and 4.11.

(d) Derive asymptotic expressions for k_1, k_2, k_3 and k_4 when $\tau \to 0$. Discuss the results.

(e) What are the values of k_1, k_2, k_3 and k_4 when $\omega_e = 0$ and U is finite? What does this say about the wavelength of the wave system along the track of a three-dimensional steady source with mean forward speed in deep water and with free-surface effects?

5 SECOND-ORDER NON-LINEAR PROBLEMS

The most common way to solve non-linear wave-structure problems in ship and offshore hydrodynamics is to use perturbation analysis with the wave amplitude as a small parameter. Potential theory is assumed and the problem is solved to second-order in incident wave amplitude. This is a powerful method that gives a solution for several practical problems.

The first-order solution is the linear solution which we have described in chapters 2, 3 and 4. In the linear solution, both the free-surface condition and the body boundary condition are satisfied on the mean position of the free-surface and the submerged hull surface respectively. Further, the fluid pressure and the velocity of fluid particles on the free-surface are linearized. What we do in a second-order theory is to account more properly for the zero-normal flow condition through the body at the instantaneous position of the body, to approximate more accurately the fluid pressure being equal to atmospheric pressure on the instantaneous position of the free-surface and to account more properly for non-linearities in the velocity of fluid particles on the free-surface. However, we are not solving the problem exactly. In a second-order theory we keep all terms in the velocity potential, fluid pressure and wave loads that are either linear with the wave amplitude or proportional to the square of the wave amplitude. The solution of the second-order problem results in mean forces, and forces oscillating with difference frequency and sum frequencies in addition to the linear solution. With a difference or a sum frequency we mean either the difference or the sum of two frequencies used in describing the wave spectrum.

Mean and slowly-varying wave loads (difference frequency loads) are of importance in several contexts for marine structures. Examples are in the design of mooring and thruster systems, analysis of offshore loading systems, evaluation of towing of large gravity platforms from the fabrication site to the operation site, added resistance of ships in waves, performance of submarines close to the free-surface and analysis of slowly oscillating heave, pitch and roll of large-volume structures with low waterplane area. Sum frequency forces can excite resonant oscillations of TLPs in heave, pitch and roll. This is referred to as 'springing' in the literature and can contribute to fatigue of tethers.

The order of magnitude of mean wave forces relative to linear first-order wave forces can be exemplified by results presented by Zhao *et al.* (1988) for a hemisphere floating in the free-surface. Incident regular waves with wave amplitude ζ_a were studied in a limited frequency range. By translating the results to a full-scale diameter of 50 m we find the linear wave excitation forces to be about $10^4\zeta_a$ (kN) while the horizontal drift force is about $10^2\zeta_a^2$ (kN). For a wave amplitude of 1 m the linear wave excitation forces are 100 times larger, and for a wave amplitude of 10 m the first-order force is about 10 times larger, than the mean wave forces. Since mean wave forces and moments are small, a high degree of accuracy is needed both in the calculations and in the experiments. The results are sensitive to wave heading, body form, body motion, wavelength and wave height. In irregular seas the second-order loads are also sensitive to the width of the wave spectrum, i.e. the wave frequency range with significant wave energy.

In a potential flow model wave drift forces are due to a structure's ability to cause waves. The consequence of this is that drift forces are small in a potential flow model when mass forces dominate. This occurs for semi-submersibles. However, viscous effects may also contribute to drift forces. This is a third-order effect that will briefly be discussed later in the chapter. Let us try to give a simple explanation of why we get a mean wave force on a structure in regular incident harmonically oscillating waves. For a surface piercing body a major contribution to the horizontal mean wave force is due to the relative vertical motion between the structure and the waves. This causes some of the body surface to be part of the time in the water and part of the time out of the water. Examining the pressure in one of the points in this surface zone, it is obvious from Fig. 5.1 that the result is a non-zero mean pressure even in regular harmonically oscillating waves. If the relative vertical motion differs around the waterline, the result is a non-zero mean force. This occurs for large-volume structures where the incident waves are modified by the structure. However, there are also other contributions. We get one of them by averaging the quadratic term in Bernoulli's equation.

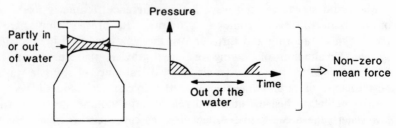

Fig. 5.1. Horizontal mean wave force contribution due to pressure forces on the free-surface zone of a structure.

A simple way to illustrate the presence of non-linear wave effects is to consider the quadratic velocity term in Bernoulli's equation for the fluid pressure. We can write this term as

$$-\frac{\rho}{2}(V_1{}^2 + V_2{}^2 + V_3{}^2) = -\frac{\rho}{2}|\nabla\phi|^2 \tag{5.1}$$

where $\mathbf{V} = (V_1, V_2, V_3)$ is the fluid velocity vector. We emphasize that equation (5.1) provides only one of the non-linear effects. Other contributions may be equally important.

Consider an idealized sea state consisting of two wave components of circular frequencies ω_1 and ω_2. An approximation for the x-component of the velocity can be written formally as

$$V_1 = A_1 \cos(\omega_1 t + \epsilon_1) + A_2 \cos(\omega_2 t + \epsilon_2) \tag{5.2}$$

By algebra it now follows that

$$
\begin{aligned}
-\frac{\rho}{2}V_1{}^2 = -\frac{\rho}{2}\Bigg[&\frac{A_1{}^2}{2} + \frac{A_2{}^2}{2} + \frac{A_1{}^2}{2}\cos(2\omega_1 t + 2\epsilon_1) \\
&+ \frac{A_2{}^2}{2}\cos(2\omega_2 t + 2\epsilon_2) \\
&+ A_1 A_2 \cos[(\omega_1 - \omega_2)t + \epsilon_1 - \epsilon_2] \\
&+ A_1 A_2 \cos[(\omega_1 + \omega_2)t + \epsilon_1 + \epsilon_2]\Bigg]
\end{aligned}
\tag{5.3}
$$

This means that we have the presence of a constant term represented by $-0.5\rho\,(A_1{}^2/2 + A_2{}^2/2)$ and a pressure term which oscillates with the difference frequency $\omega_1 - \omega_2$. For a more realistic representation of the seaway, and considering the wave as the sum of N components of different circular frequencies ω_i, we will find pressure terms with difference frequencies $\omega_j - \omega_k$, $(k, j = 1, N)$. These non-linear interaction terms produce slowly-varying excitation forces and moments which may cause resonance oscillations in the surge, sway, and yaw motions of a moored structure. Typical resonance periods are 1–2 minutes.

Equation (5.3) also tells us that non-linear effects can create excitation forces with frequencies higher than the dominant frequency components in a wave spectrum. This is due to terms oscillating with frequencies $2\omega_1$, $2\omega_2$ and $(\omega_1 + \omega_2)$. These may be important for exciting the resonance oscillations in the heave, pitch and roll for TLPs. Typical resonance periods are 2–4 s. The restoring forces for the TLP are due to the tethers and the mass forces due to the structure. In the following text we will start by discussing mean wave loads.

MEAN WAVE (DRIFT) FORCES AND MOMENTS

When calculating the mean wave loads on a structure it is not necessary to solve the second-order problem. We will explain this in the case where there is no current and the structure has no forward speed. We will assume incident regular waves with amplitude ζ_a and circular frequency ω. Let ϕ_2 be the velocity potential for the second-order potential. This means ϕ_2 is proportional to the square of the incident wave amplitude. We can show that the time dependence of ϕ_2 can be expressed as $A + B \cos(2\omega t + \epsilon)$, where A and B are independent of time. The reason is that the second-order problem involves solving a boundary value problem where the boundary conditions are expressed as the product of two terms. Each of the terms is from the first-order solution and are harmonically oscillating with frequency ω. The product of the terms will give one term that is time independent and one term that is oscillating with frequency 2ω. The second-order potential will have the same time dependence as the boundary conditions for the problem. The pressure associated with the second-order potential can be found from Bernoulli's equation. By keeping terms that are proportional to ζ_a^2 the pressure can be written $-\rho\, \partial\phi/\partial t = \rho 2\omega B \sin(2\omega t + \epsilon)$. The mean value over one period of oscillation of this pressure part is zero. This means the second-order potential does not result in mean wave loads. All information that we need can be found from the linear first-order solution. Results in an irregular sea can be obtained by adding results from regular waves. We will therefore start with discussing the effect in regular waves.

One way to obtain expressions for mean wave forces in regular waves is to use the equations for conservation of momentum $\mathbf{M}(t)$ in the fluid. Let S be a closed surface. The momentum inside S can be written as

$$\mathbf{M}(t) = \iiint_{\Omega} \rho \mathbf{V}\, d\tau \tag{5.4}$$

where $\mathbf{V} = (V_1, V_2, V_3)$ is the fluid velocity. The enclosing surface S does not need to follow the fluid motion. By using the definition of a derivative and noting that both the volume and the velocity may change with time, it follows that (see Fig. 5.2)

$$\frac{d\mathbf{M}}{dt} = \rho \iiint_{\Omega} \frac{\partial \mathbf{V}}{\partial t}\, d\tau + \rho \iint_{S} \mathbf{V} U_n\, ds \tag{5.5}$$

Here U_n is the normal component of the velocity of the surface S. Note that we have defined here the positive normal direction to be out of the fluid. The last integral is the effect of integrating over the shaded area in Fig. 5.2 and letting Δt be small (i.e. go to zero). The volume integral in equation (5.5) can be rewritten by expressing $\partial \mathbf{V}/\partial t$ by Euler's equation.

This can, for an incompressible fluid, be written as

$$\frac{\partial \mathbf{V}}{\partial t} + \mathbf{V} \cdot \nabla \mathbf{V} = -\nabla \left(\frac{p}{\rho} + gz \right)$$

The volume integral can be reduced to a surface integral by using vector algebra and a generalized Gauss theorem which states that

$$\iiint\limits_{\Omega} \nabla \circ X \, d\tau = \iint\limits_{S} \mathbf{n} \circ X \, ds \qquad (5.6)$$

Here X may be a scalar, vector or tensor and $\circ$ denotes a dot or cross or an ordinary multiplication or nothing. It is assumed that X has continuous derivatives in Ω. We can then show that

$$\frac{d\mathbf{M}}{dt} = -\rho \iint\limits_{S} \left[\left(\frac{p}{\rho} + gz \right) \mathbf{n} + \mathbf{V}(V_n - U_n) \right] ds \qquad (5.7)$$

Here $V_n = \partial \phi / \partial n$ is the normal component of the fluid velocity at the surface S.

This is a general formula which we will apply to the wave-drift force problem. We let the closed surface S consist of the body surface S_B, a non-moving vertical circular cylindrical surface S_∞ away from the body, the free-surface S_F and the sea bottom S_0 inside S_∞ (see Fig. 5.3). We should note that S_∞ does not need to be far away from the body. We can write

$$U_n = V_n \quad \text{on } S_B \text{ and } S_F$$

$$U_n = 0 \quad \text{on } S_\infty \text{ and } S_0 \qquad (5.8)$$

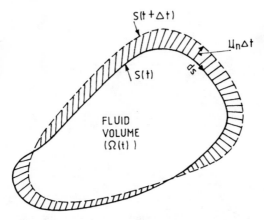

Fig. 5.2. Illustration of how the control volume $\Omega(t)$ changes in a time increment Δt. (U_n = normal component of the velocity of the surface S).

The force $\mathbf{F} = (F_1, F_2, F_3)$ on the body is $\int_{S_B} p\mathbf{n}\,ds$. On the free-surface the pressure is equal to the atmospheric pressure p_0. However, since $\iint_S p_0\mathbf{n}\,ds = 0$ we could just as well define p in equation (5.7) to be the difference between pressure in the fluid and the atmospheric pressure. This implies that we can set $p = 0$ on S_F. The term $\rho \iint_S gz\mathbf{n}\,ds$ gives no horizontal component. By time averaging equation (5.7) over one period of oscillation and noting that the time average of $d\mathbf{M}/dt$ is zero, we find that

$$\bar{F}_i = -\overline{\iint_{S_\infty} [pn_i + \rho V_i V_n]\,ds} \qquad i = 1,2 \tag{5.9}$$

When calculating $\bar{F}_i$ by equation (5.9) we can use Bernoulli's equation to calculate the pressure. We should realize that the wetted area of S_∞ is time-dependent due to the time-varying wave elevation at S_∞. In evaluating equation (5.9) correctly to second-order in wave amplitude, it is only necessary to know the first-order velocity potential (see the discussion in the beginning of the chapter). The first-order potential can for instance be calculated by the numerical methods described in chapter 4. If we want to use equation (5.7) for conservation of momentum to calculate the mean vertical force, we note from equation (5.7) that we do not get rid of the ρgz-term in the integrand and have to include this contribution from an integral over the free-surface.

Equation (5.9) was derived by Maruo (1960). Newman (1967) derived a similar formula for the mean wave-drift yaw moment. He started out by using the fluid angular momentum

$$\mathbf{K}(t) = \rho \iiint_\Omega (\mathbf{r} \times \mathbf{V})\,d\tau \tag{5.10}$$

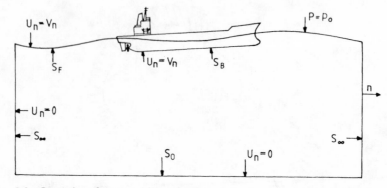

Fig. 5.3. Control surfaces used in evaluating expressions for mean wave loads and illustrations of how the normal velocity U_n of the control surfaces relates to the normal velocity of the fluid.

where **r** is the position vector relative to the origin of the coordinate system (x, y, z), which is fixed in space. The derivation is similar to that shown for equation (5.7). If we define n_6 and V_6 from $\mathbf{r} \times \mathbf{V} = (V_4, V_5, V_6)$ and $\mathbf{r} \times \mathbf{n} = (n_4, n_5, n_6)$, the result is that equation (5.9) is also valid for the yaw moment with $i = 6$.

Equation (5.9) also applies in two dimensions. In that case we write S_∞ for the sum of S_∞ and $S_{-\infty}$, where S_∞ and $S_{-\infty}$ are vertical control surfaces for positive and negative horizontal coordinates respectively. The control surface can be any distance away from the body even if the subscript '∞' indicates that the control surfaces S_∞ and $S_{-\infty}$ are far away from the body.

Maruo's formula

Maruo (1960) used equation (5.9) to derive a useful formula for drift forces on a two-dimensional body in incident regular deep-water waves. The body may be fixed or freely floating oscillating around a mean position. There is no current and the body has no constant speed.

The velocity potential for the incident waves is written as (see Fig. 5.4)

$$\phi_0 = \frac{g\zeta_a}{\omega} e^{kz} \cos(\omega t - ky) \tag{5.11}$$

with ζ_a being the amplitude of the waves. Due to the presence of the body there will be generated waves. The velocity potential of the

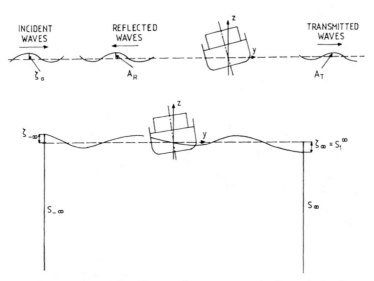

Fig. 5.4. Definition of control surfaces and wave systems in the analysis of drift forces on a two-dimensional body.

reflected waves is written as

$$\phi_R = \frac{gA_R}{\omega} e^{kz} \cos(\omega t + ky + \epsilon) \tag{5.12}$$

where the wave amplitude A_R and phase angle ϵ are functions of the body form and motion. The velocity potential of the transmitted waves is written as

$$\phi_T = \frac{gA_T}{\omega} e^{kz} \cos(\omega t - ky + \delta) \tag{5.13}$$

ϕ_T is a combination of the incident waves and the waves generated at $y \rightarrow \infty$ by the body.

When the two-dimensional version of equation (5.9) is used to find the horizontal drift-force $\bar{F}_2$, it is convenient to divide S_∞ into two parts. One part S_1^∞ is from the mean free-surface level $z = 0$ to the instantaneous position ζ_∞ of the free-surface (see Fig. 5.4). Since ζ_∞ is a first-order quantity and we are evaluating the integral correctly to second order in wave amplitude, it is only necessary to evaluate the integrand correctly to first order when we integrate along S_1^∞. This means the term ρV_2^2 can be neglected and the pressure p approximated by

$$- \rho g z - \rho \frac{\partial \phi_T}{\partial t} \bigg|_{z=0}$$

along S_1^∞. The result is a contribution $-\frac{1}{2}\rho g \overline{\zeta_\infty^2} = -\frac{1}{4}\rho g A_T^2$.

In the integration from $z = -\infty$ to $z = 0$ we have to keep second-order terms in the integrand. The linear terms give a zero mean value. The hydrostatic pressure $-\rho g z$ will result in a force contribution that is balanced out by a similar contribution at the control surface at $y = -\infty$. The integration along $S_{-\infty}$ is performed in a similar way as the integration over S_∞. The total result is

$$\bar{F}_2 = \frac{\rho g}{4} [\zeta_a^2 + A_R^2 - A_T^2] \tag{5.14}$$

Longuet-Higgins (1977) has generalized equation (5.14) to finite water depth. The right hand side should then be multiplied by $(1 + 2kh/\sinh 2kh)$, where h is the water depth.

Maruo's formula follows by assuming the average energy flux is zero through S_B. This means

$$\zeta_a^2 = A_R^2 + A_T^2 \tag{5.15}$$

We can show this by using equation (3.21) and noting that there is no work done on the body during one period of oscillation. This means $\iint_{S_B} (p - p_0) U_n \, ds = 0$. By using $\overline{dE/dt} = 0$, it follows from equation

(3.22) that $\overline{\iint_{S_\infty + S_{-\infty}} (\partial\phi/\partial t)(\partial\phi/\partial n)\, ds} = 0$. From this condition we can show equation (5.15) by using equations (5.11), (5.12) and (5.13). Equation (5.14) can then be written as

$$\bar{F}_2 = \frac{\rho g}{2} A_R{}^2 \tag{5.16}$$

According to the formula the wave-drift force will always act in the wave propagation direction. Due to condition (5.15) Maruo's formula (equation (5.16)) is not valid if the body is an active wave power device.

Equation (5.16) shows that wave-drift forces are connected with a structure's ability to cause waves. The waves due to the body are the sum of (a) the radiating waves when the body is forced to oscillate in each mode of motion, and (b) the diffraction waves when the body is restrained from oscillating and subject to incident waves. For long wavelengths relative to the cross-sectional dimensions the body will not disturb the wave field. This means the reflected wave amplitude A_R and the wave drift force become negligible.

When the wavelengths are very short, the incident waves are totally reflected from a surface-piercing body with vertical hull surface in the wave zone. This means $A_R = \zeta_a$ and $\bar{F}_2 = (\rho g/2)\zeta_a{}^2$. According to equation (5.15) the reflected wave amplitude can never be larger than ζ_a. This means the wave-drift force can never be larger than $(\rho g/2)\zeta_a{}^2$.

When the body motions are large, for instance due to heave resonance, A_R is likely to be large. This means that the wave-drift force may show a peak in a frequency range around the resonance frequency. This can be seen in Fig. 5.5.

For a submerged body A_R will go to zero when the wavelength goes to zero. In the special case of a submerged circular cylinder that is either restrained from oscillating or whose centre follows a circular orbit Ogilvie (1963) showed that A_R is zero for all frequencies and all depths of submergence.

The order of magnitude of wave-drift forces $\bar{F}_2$ on a free-surface-piercing two-dimensional body relative to current forces $\bar{F}_c$ on the same structure when there are no waves present can be estimated by

$$\frac{\bar{F}_2}{F_c} = \frac{0.5\rho g A_R{}^2}{0.5\rho C_D D U_c{}^2} = \frac{g A_R{}^2}{C_D D U_c{}^2}$$

Here C_D is the drag coefficient for the current flow, D is the draught and U_c is the current velocity. For instance, in a current velocity of 1 m s^{-1}, a draught $D = 10$ m and $C_D = 1.0$ the ratio is simply $A_R{}^2$ in square metres. This illustrates the importance of wave-drift forces on large volume structures like ships. However, one must not be misled by this discussion into believing that the effect of current and waves can be superposed. The combined effect of waves and current have an effect on the wave field and therefore on the wave-drift forces. Further, the combined effect of waves and current influences the occurrence of flow-separation and the drag coefficient C_D.

Maruo (1960) has also derived a formula similar to equation (5.16) for drift-forces on a three-dimensional structure in incident regular waves, with no current present, which can be written in the form

$$\bar{F}_1 = \frac{\rho g}{4} \int_0^{2\pi} A^2(\theta)(\cos \beta - \cos \theta)\, \mathrm{d}\theta \qquad (5.17)$$

$$\bar{F}_2 = \frac{\rho g}{4} \int_0^{2\pi} A^2(\theta)(\sin \beta - \sin \theta)\, \mathrm{d}\theta \qquad (5.18)$$

Here β is the wave propagation direction relative to the x-axis and $A(\theta)/r^{\frac{1}{2}}$ is the wave amplitude generated by the body far away at large horizontal radial distance $r = (x^2 + y^2)^{\frac{1}{2}}$ from the body. These waves are the sum of radiation waves when the body is oscillating in six degrees of freedom and the diffraction waves when the body is restrained from oscillating and subject to incident waves. The angle θ is defined as $x = r \cos \theta$, $y = r \sin \theta$. Equations (5.17) and (5.18) are derived on the

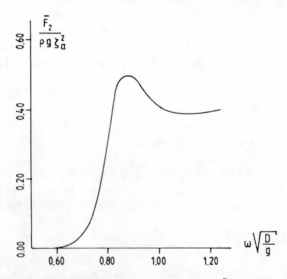

Fig. 5.5. Example of behaviour of wave drift force $\bar{F}_2$ on a two-dimensional free-surface piercing body Lewis form section with $B/D = 2.0$, $A = 0.95BD$. Beam seas waves. Infinite water depth. (B = Beam, D = Draught, A = submerged cross-sectional area, ζ_a = wave amplitude of the incident waves, ω = circular frequency of oscillations.)

assumption that energy is conserved. For example, if viscous forces matter these expressions are not valid. Equations (5.17) and (5.18) also demonstrate that wave drift forces in a potential flow model are due to a structure's ability to cause waves. If we apply equations (5.17) and (5.18) to head sea waves incident on a ship, we find that the wave-drift force is always in the wave propagation direction. For a general wave heading, the resulting drift force may not necessarily coincide with the wave propagation direction. We cannot see from equations (5.17) and (5.18) what wave directions give largest wave-drift forces on a ship. To illustrate the wave heading dependence of drift forces let us present two examples of calculations by Faltinsen *et al.* (1979) (see Fig. 5.6). For the smallest wavelength the transverse drift force is largest for beam sea waves, while for the larger wavelength it is practically zero for beam sea and largest for wave headings around 45° and 135°. The reason is that the ship follows the waves best at 90° so that the generated wave amplitudes are small.

By interference between waves generated from different parts of the structure, the generated waves may become small for certain wavelengths and headings and result in small drift forces. This is typical for a TLP.

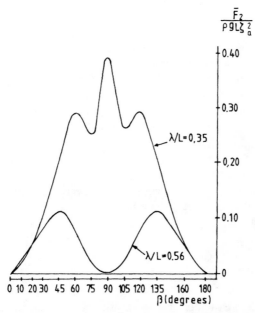

Fig. 5.6. Transverse drift force $\bar{F}_2$ on a ship as a function of wave heading β for two wavelengths λ. Infinite water depth. (L = ship length, $\beta = 0°$ in head seas, ζ_a = wave amplitude of the incident waves.) (Faltinsen *et al.*, 1979.)

As a first approximation one may imagine the columns of a TLP as independent wave generators. Dependent on the wavelength and heading the wave fields generated by each column may either tend to enforce or cancel each other.

If a roll resonance occurs, equations (5.16), (5.17) and (5.18) are not expected to give satisfactory results for ship hulls. The reason is that viscous effects are important in calculating roll resonance amplitudes (see chapter 3) whereas the formula is based on potential flow theory.

Direct pressure integration method

Another way to obtain mean wave forces and moments is to use the direct pressure integration method (Pinkster & van Oortmerssen, 1977). This means one starts out with Bernoulli's equation for the pressure and formally writes forces and moments on the hull correctly to second order in wave amplitude. All three force components and three moment components can be obtained. However, to avoid inaccuracies care should be taken in using the method. For instance, if a boundary element method is used to calculate the linear flow motion and one assumes constant velocity potential over each element, one has implicitly said that the tangential velocity is zero over the element. This means error in the calculation of the quadratic velocity term in Bernoulli's equation, in particular close to sharp corners. There are ways to avoid this difficulty, for instance by extrapolating calculated values of fluid velocities from points along the normal to the element that are further away than a characteristic length of the element.

We will exemplify the method by analysing incident regular deep-water waves on a two-dimensional free-surface-piercing body. We assume the wavelength is very small and that the surface of the cylinder is vertical at the intersection with the free-surface. Since the wavelength is small, the cylinder will not oscillate in the waves. Further, the effect of the waves is only felt in the free-surface zone on the 'upstream' side. On the 'downstream' side there is a shadow region. From a hydrodynamic point of view this means that we could just as well study incident waves on a vertical wall (see Fig. 5.7). The linear solution of this problem can be written

$$\phi_1 = \frac{2g\zeta_a}{\omega} e^{kz} \cos \omega t \cos ky$$

Physically this represents standing waves. The maximum wave elevation at the wall is twice the incident wave amplitude. If we calculate the linear hydrodynamic force, we find

$$F_1 = -\int_{-\infty}^{0} \rho \frac{\partial \phi_1}{\partial t}\bigg|_{y=0} dz = 2\rho g \zeta_a \frac{1}{k} \sin \omega t$$

This is a harmonically oscillating force, which cannot explain the drift-forces. We have to include higher-order terms in our analysis. This means we must integrate over the exact wetted surface of the body and use the complete Bernoulli's equation. Further, we have to solve the hydrodynamic problem to higher-order in wave amplitude. However, it can be shown as discussed earlier that the contribution from the second-order potential to the mean force is zero in regular waves. Let us start out with Bernoulli's equation

$$p = -\rho g z - \rho \frac{\partial \phi_1}{\partial t} - \frac{\rho}{2}\left(\left(\frac{\partial \phi_1}{\partial y}\right)^2 + \left(\frac{\partial \phi_1}{\partial z}\right)^2\right) \qquad (5.19)$$

We will calculate the force correct to second order in wave amplitude and exclude the hydrostatic force.

The contribution from the two first terms in equation (5.19) can be written as

$$\overline{-\rho g \int_0^\zeta z \, dz} - \overline{\rho \frac{\partial \phi_1}{\partial t}\bigg|_{z=0} \zeta} = \rho g \zeta_a^2 \qquad (5.20)$$

Here $\zeta = 2\zeta_a \sin \omega t$ is the wave amplitude at the wall. The third term in equation (5.19) results in the following contribution

$$\overline{-\frac{\rho}{2}\int_{-\infty}^0 \left\{\left(\frac{\partial \phi_1}{\partial y}\right)^2 + \left(\frac{\partial \phi_1}{\partial z}\right)^2\right\} dz}$$

$$= -\frac{\rho}{2}\int_{-\infty}^0 \frac{1}{2}\frac{4g^2\zeta_a^2}{\omega^2} k^2 e^{2kz} \, dz = -\tfrac{1}{2}\rho g \zeta_a^2$$

The total sum is $(\rho g/2)\zeta_a^2$. This is the correct asymptotic value for small wavelengths according to Maruo's formula. We can generalize the asymptotic formula to any structure as long as the structure has vertical

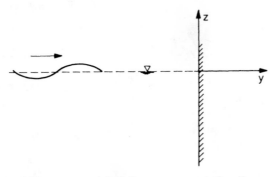

Fig. 5.7. Incident waves and drift forces on a vertical wall.

sides at the waterplane. We can write

$$\bar{F}_i = \frac{\rho g \zeta_a^2}{2} \int_{L_1} \sin^2(\theta + \beta) n_i \, dl \tag{5.21}$$

Here $\bar{F}_1$ is the drift-force component in the x-direction, $\bar{F}_2$ is the drift-force component in the y-direction and $\bar{F}_6$ is the yaw-drift moment with respect to the z-axis (see Fig. 5.8 for definitions).

Further

$$n_1 = \sin \theta$$

$$n_2 = \cos \theta$$

$$n_6 = x_0 \cos \theta - y_0 \sin \theta$$

The integration in equation (5.21) is along the non-shadow part L_1 of the waterplane curve, i.e. from A to B. Further, β is the wave propagation direction and the angle θ is defined in Fig. 5.8.

We have worked out three special cases below

Example 1
Infinitely long horizontal cylinder

$$\bar{F}_1 = 0, \qquad \bar{F}_2 = \frac{\rho g}{2} \zeta_a^2 \sin \beta \, |\sin \beta|, \qquad \bar{F}_6 = 0 \, (\text{per unit length})$$

Example 2
Structure with circular waterplane area of radius r

$$\bar{F}_1 = \tfrac{2}{3} \rho g \zeta_a^2 r \cos \beta, \qquad \bar{F}_2 = \tfrac{2}{3} \rho g \zeta_a^2 r \sin \beta, \qquad \bar{F}_6 = 0$$

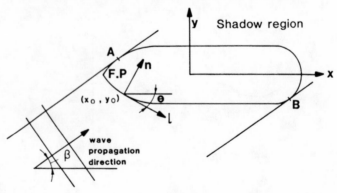

Fig. 5.8. Definition of ship and wave parameters used in equation (5.21).

Example 3

Structure with waterplane area consisting of two circular ends of radius r and a parallel part of length $2l$ (see Fig. 5.9). We find that

$$\bar{F}_1 = \tfrac{2}{3}\rho g \zeta_a^2 r \cos \beta, \qquad \bar{F}_2 = \rho g \zeta_a^2 (\tfrac{2}{3} r \sin \beta + l \sin \beta \, |\sin \beta|)$$

$$\bar{F}_6 = -\frac{\rho g \zeta_a^2}{3} l r \sin 2\beta$$

It should be noted that the angular dependences shown in the formula above are, strictly speaking, only valid for small wavelengths. For other wavelengths quite different angular dependence may occur (Faltinsen *et al.*, 1979).

Added resistance of ships in waves

Equation (5.21) does not take into account current or the forward motion of a structure. Faltinsen *et al.* (1980) have derived a formula for added resistance, R_{AW}, of a ship in small wavelengths. Added resistance is the same as the longitudinal drift-force component. The formula is assumed to be valid for small Froude numbers, i.e. $F_n < \approx 0.2$, and blunt ship forms. All wave headings are considered. For head sea we can write

$$\frac{\bar{F}_1}{\zeta_a^2} = \tfrac{1}{2}\rho g \left(1 + \frac{2\omega_0 U}{g}\right) \int_{L_1} \sin^2 \theta \, n_1 \, dl \qquad (5.22)$$

Here ω_0 is the circular frequency of oscillation of the waves and U is the forward speed of the ship. L_1 is the non-shadow part of the waterplane curve (see Fig. 5.8). The formula is not restricted to a ship and may for instance be used in connection with towing of marine structures in waves. Equation (5.22) is sensitive to the geometric form in the bow region. This is illustrated in Fig. 5.10 for a ship, where F_0 means a reference value for the added resistance.

To illustrate the effect of forward speed, we see that equation (5.22)

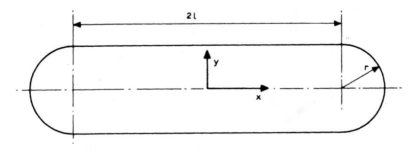

Fig. 5.9. Definition of parameters.

shows that

$$\frac{\text{Added resistance }(Fn \neq 0)}{\text{Added resistance }(Fn = 0)} = 1 + \frac{2\omega_0 U}{g}$$

If we set $\lambda/L = 0.5$, the ratio becomes $1 + 7 \cdot Fn$, which means a strong forward speed dependence. The formula also indicates that the effect of current cannot be neglected. This was confirmed by Zhao et al. (1988); who pointed out that current velocities of 1 m s^{-1} may very well represent a 50% increase in drift forces on large volume structures. This was based on both numerical and experimental investigations. Qualitatively one can understand that current must have an effect on mean wave forces if one considers that drift forces are due to a structure's ability to cause waves and the fact that current has an influence on the wave field (see Fig. 4.13).

A typical curve for the added resistance R_{AW} of a ship at a moderate Froude number in head sea regular waves is presented in Fig. 5.11. There are two main features in the figure.

(a) For small wavelengths ($\lambda/L < 0.5$) the added resistance is mainly due to reflection of waves from the bow of the ship. This effect is described by equation (5.22).

(b) For large wavelengths the added resistance has a maximum when λ is of the order of the ship length. This maximum occurs when the relative vertical motion between the ship and the waves is large. For moderate Froude numbers below 0.3 the maximum value of $\sigma_{AW} = R_{AW}/(\rho g \zeta_a^2 B^2/L)$ is normally below 20. For a high speed hull form at $Fn = 0.8$ Strom-Tejsen et al. (1973) reported values of σ_{AW} close to 35.

Gerritsma & Beukelman (1972) have derived a widely-used formula for added resistance of a ship in regular waves. The formula is based on a

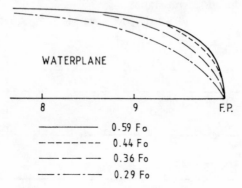

WATERPLANE

8   9   F.P.

———————— 0.59 Fo
– – – – – – – 0.44 Fo
—— —— —— 0.36 Fo
—— · —— · —— 0.29 Fo

Fig. 5.10. Influence of bow form on added resistance for small wave lengths. (F_0 = reference value for added resistance.) Head sea waves.

strip-theory approximation and shows that

$$R_{\mathrm{AW}} = \frac{k}{2\omega_e} \int_L \left(B_{33}^{(2D)} + U \frac{\mathrm{d}}{\mathrm{d}x} A_{33}^{(2D)} \right) V_{za}^2(x)\, \mathrm{d}x \qquad (5.23)$$

The integration is along the length of the ship and $V_{za}(x)$ is the amplitude of the relative vertical velocity between the ship and the waves. The formula demonstrates that added resistance depends strongly on the relative vertical motion between the ship and the waves. The quality of the predictions by the formula is sensitive to the accuracy of the calculation of relative vertical velocity. The formula should be used with care for unconventional ship forms and fishing vessels. It is also questionable in the small wavelength range for blunt ship forms. One reason is that the formula is based on slender body (strip) theory and neglects the effect of reflection of waves from the bow of the ship.

We can obtain a crude approximation for the added resistance of a ship in the 'ship-motion' range by starting out with a formula proposed by Faltinsen et al. (1980). This formula was derived by a direct pressure integration approach and is valid for regular sinusoidal incident waves. When the added resistance is at its maximum, a dominant contribution in the formula is the relative motion term

$$\bar{F}_1 \approx -\frac{\rho g}{2} \int_c \overline{\zeta_R^2 n_1}\, ds$$

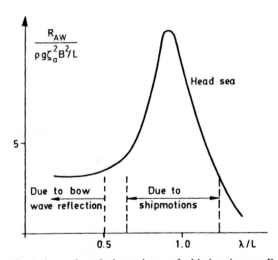

Fig. 5.11. Typical wavelength dependence of added resistance R_{AW} of a ship at moderate Froude number in regular head sea waves. (ζ_a = wave amplitude of incident waves, λ = wavelength, L = ship length, B = beam of the ship.)

where the integration is with respect to the waterline curve, c, n_1 is the longitudinal component of the outward normal to c, ζ_R is the relative vertical motion along c.

In the case of a ship with a constant cross-section, we can re-express this formula in the form

$$\bar{F}_1 \simeq \frac{\rho g}{2} B(\overline{\zeta_{RB}^2} - \overline{\zeta_{RA}^2}) \tag{5.24}$$

where B is the beam of the ship and ζ_{RB} and ζ_{RA} are the relative vertical motions at the bow and the stern respectively. This formula does not imply that the added resistance is linearly dependent on B, since the relative motion ζ_R is also dependent on B (see discussion in chapter 3). Equation (5.24) can be used to discuss how the added resistance in the 'ship motion' range depends on main hull characteristics (see exercise 5.3).

Irregular sea calculations

When the results of mean wave loads in regular waves are known, it is easy to obtain results in an irregular sea. We will show a formula for the loads. We assume long-crested seas that can be described by a sea spectrum $S(\omega)$. In deep water we can write the velocity potential of the incident waves correctly to first order in wave amplitude as

$$\phi^I = \sum_{j=1}^{N} \frac{gA_j}{\omega_j} e^{k_j z} \cos(\omega_j t - k_j x \cos \beta - k_j y \sin \beta + \epsilon_j) \tag{5.25}$$

Here β is the angle between the wave propagation direction and the x-axis, ω_j the circular frequency of oscillation and k_j the wave number of wave component number j, ω_j and k_j are connected through the dispersion relationship $\omega_j^2/g = k_j$. The phase angles ϵ_j may be considered as random phase angles and the amplitudes A_j may be determined by the wave spectrum $S(\omega)$ characterizing the sea state. If the major part of the wave energy is concentrated between the circular frequencies ω_{min} to ω_{max}, we divide the frequency interval ω_{min} to ω_{max} into N equal sub-intervals and call the mid-points of the jth interval ω_j (see Fig. 5.12). A_j is then determined by

$$\frac{A_j^2}{2} = S(\omega_j) \frac{\omega_{max} - \omega_{min}}{N} \tag{5.26}$$

In principle we should let $N \to \infty$, $\omega_{min} \to 0$ and $\omega_{max} \to \infty$ so that the sum in equation (5.25) becomes an integral. If N is kept finite and we want to use equation (5.25) to simulate an irregular sea, the signal repeats itself after a period $2\pi N/(\omega_{max} - \omega_{min})$. To avoid this we can choose ω_j randomly in each sub-interval $\Delta\omega_j$ in Fig. 5.12.

Let us now imagine that we want to find the mean wave loads on the structure by using the direct pressure integration method. One part of the procedure is to study the effect of the quadratic velocity term in Bernoulli's equation, i.e. equation (5.1). In equations (5.2) and (5.3) we studied the consequence of using an idealized sea state with two wave components. The results in equation (5.3) show that we can linearly *add* together the mean force contribution from each wave component. This may be surprising since we talk about a system that is non-linear in the wave amplitude. The same would have happened if we had used N wave components. Similar summation can be done for the other contributions to the mean wave loads. What we now say is that the mean wave load can be written as

$$\overline{F_i^s} = \sum_{j=1}^{N} \left(\frac{\overline{F_i}(\omega_j, \beta)}{\zeta_a^2} \right) A_j^2 \qquad i = 1, \ldots, 6 \qquad (5.27)$$

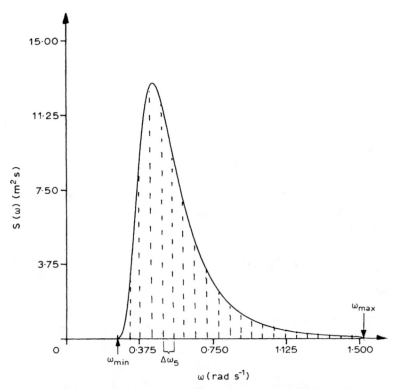

Fig. 5.12. Illustration of how the energy in a wave spectrum $S(\omega)$ can be distributed into energy of regular wave components of circular frequency ω. (ISSC-spectrum, $H_{\frac{1}{3}} = 8$ m, $T_2 = 10$ s, number of wave components $N = 22$.) See also Fig. 2.5.

Here $\bar{F}_i(\omega_j; \beta)$ is the ith mean wave load component in regular incident waves of circular frequency ω_j, wave amplitude A_j and wave propagation direction β. Further, $\bar{F}_i(\omega_j; \beta)$ is divided by the wave amplitude square, i.e. ζ_a^2, so that $\bar{F}_i(\omega_j; \beta)/\zeta_a^2$ is independent of the wave amplitude. In integral form we can write equation (5.27) as

$$\overline{F_i^s} = 2\int_0^\infty S(\omega)\left(\frac{\bar{F}_i(\omega; \beta)}{\zeta_a^2}\right) d\omega \qquad i = 1, \ldots, 6 \qquad (5.28)$$

Let us illustrate the use of equation (5.28) by applying the equation to the added resistance of a ship in head sea waves and use equation (5.22) to calculate $\bar{F}_1/\zeta_a^2$. This means we assume the wavelengths of the wave components of practical importance are small relative to the ship length. By using equation (5.22) we will be able to integrate equation (5.28) analytically. However, this sets certain limitations on the form of the sea spectrum. We must require there to be no significant wave energy for wavelengths larger than half the ship length. For a Pierson–Moskowitz spectrum with two parameters $H_{\frac{1}{3}}$ and T_2 (see equation (2.24)), we can tentatively state that there is no significant wave energy for

$$\frac{\omega T_2}{2\pi} < 0.5$$

Here $H_{\frac{1}{3}}$ is the significant wave height and T_2 is the mean period defined by the second moment of the wave spectrum. By using the dispersion relationship for deep water waves and the two requirements mentioned above, we find that we can only combine equation (5.22) with equation (5.28) when

$$T_2 < 0.9(L/g)^{\frac{1}{2}} \tag{5.29}$$

For oceangoing vessels this means the ship length has to be large for the formula to be of any significant practical use.

By evaluating equation (5.28) we find that

$$\overline{F_1^s} = \rho g \frac{H_{\frac{1}{3}}^2}{16}\left(1 + \mathrm{Fn}\frac{4\pi}{T_1}\sqrt{\frac{L}{g}}\right)\int_{L_1} \sin^2\theta\, n_1\, dl \tag{5.30}$$

We have used here

$$H_{\frac{1}{3}} = 4\sqrt{m_0} \quad \text{where} \quad m_0 = \int_0^\infty S(\omega)\, d\omega$$

and $\hspace{12cm}$ (5.31)

$$T_1 = 2\pi\frac{m_0}{m_1} \quad \text{where} \quad m_1 = \int_0^\infty \omega S(\omega)\, d\omega$$

For a Pierson–Moskowitz spectrum we may write

$$T_1 = 1.086 T_2 \tag{5.32}$$

For a one-parameter Pierson–Moskowitz spectrum we should note that condition (5.29) means that $H_{\frac{1}{3}} < 0.0065L$. For a 300 m long ship this means $H_{\frac{1}{3}} < 1.95$ m. In the North Atlantic this occurs about 40% of the time. For a 100 m long ship it means $H_{\frac{1}{3}} < 0.65$ m, which occurs less than 4% of the time in the North Atlantic.

It is worth pointing out that added resistance of a ship is sensitive to the mean wave period. This is evident from the study by Strom-Tejsen et al. (1973). The reason can be seen from Fig. 5.11. The added resistance curve has a very distinct peak around heave and pitch resonance for a ship at finite Froude number. When multiplying this result with the wave spectrum and integrating over all frequencies (see equation (5.28)), the result is sensitive to how the peak period of the sea spectrum is located relative to the peak period of the added resistance curve. Type of wave spectrum will also influence the magnitude of the added resistance.

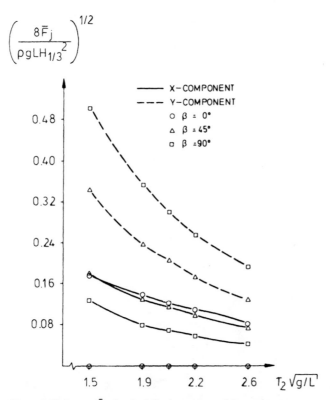

Fig. 5.13. Wave drift forces $\bar{F}_j, j = 1, 2$ for long-crested irregular waves incident on a tanker. The forward speed and current are zero. Infinite water depth. ($H_{\frac{1}{3}}$ = significant wave height, T_2 = mean wave period, L = ship length, β = wave heading with $\beta = 0°$ being head sea.)

Examples of wave drift forces and moments on a tanker in an irregular sea of different wave headings and mean wave periods are presented in non-dimensionalized form in Figs. 5.13 and 5.14. The ship speed is zero. A JONSWAP spectrum was used in the calculations. Further, $\beta = 0°$ corresponds to a head sea. A negative yaw moment means that the moment tends to rotate the ship to a position with a larger wave heading angle. The effect of mean wave period is not so pronounced in the results presented in Fig. 5.13 and 5.14 as Strom-Tejsen *et al.* (1973) showed for added resistance of a ship at forward speed.

Viscous effects on mean wave forces

When wave drift forces become small, viscous effects may contribute to drift forces. This happens for semi-submersibles when the incident wave amplitude is large relative to cross-sectional dimensions of hull com-

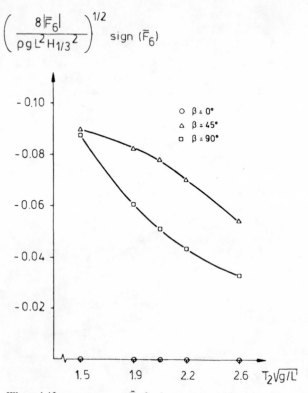

$$\left(\frac{8\,|\bar{F}_6|}{\rho g\,L^2 H_{1/3}{}^2} \right)^{1/2} \; \text{sign } (\bar{F}_6)$$

Fig. 5.14. Wave drift yaw moment $\bar{F}_6$ for long-crested irregular waves incident on a tanker. Notation as in Fig. 5.13. Positive yaw moment implies rotation of the ship towards smaller β-angles. Moment is with respect to the vertical axis through the centre of gravity of the ship. The forward speed and current are zero. Infinite water depth.

ponents. The effect is of third order which means it is proportional to the cube of the wave amplitude in regular waves. There are several reasons why viscous effects contribute to mean drift forces. By using a simple cross-flow principle for the flow around the pontoons and columns of a semi-submersible, decomposing the forces into components along an earth fixed coordinate system and averaging the forces over one wave period, we will find non-zero mean wave loads that are proportional to the third power of the wave amplitude. Let us show this in more detail. We will refer to Fig. 5.15 and write the incident wave potential as

$$\phi = \frac{g\zeta_a}{\omega} e^{kz} \cos(\omega t - kx) \tag{5.33}$$

This means the vertical incident fluid velocity can be written as

$$w = \omega\zeta_a e^{kz} \cos(\omega t - kx) \tag{5.34}$$

The drag force normal to the pontoon in the N-direction can be written as

$$F_N = -2 \int_{-L/2}^{L/2} \frac{\rho}{2} C_D b V_R |V_R|\, dx \tag{5.35}$$

Here V_R is the relative velocity in the N-direction between a strip of the platform and the incident wave field. It can be written approximately as

$$V_R \approx \frac{d\eta_3}{dt} - x\frac{d\eta_5}{dt} - w \tag{5.36}$$

This results in a longitudinal force

$$F_x = F_N \eta_5 \tag{5.37}$$

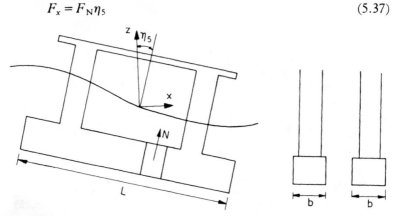

Fig. 5.15. Definition of parameters for calculation of the viscous drift force contribution from the pontoons of a semi-submersible.

Fig. 5.16 illustrates that the time average of $V_R |V_R|$ over one period is zero, while the time average of $V_R |V_R| \eta_5$ is different from zero. For simplicity we have set the phase angle of η_5 equal to the phase angle of V_R. The phase angle between η_5 and V_R will determine in what direction the mean force contribution works, i.e. in line with, or opposite to, the wave propagation direction.

The time average of F_x is one of the contributions to mean wave forces induced by drag forces. We can analyse the columns in a similar way. By evaluating the viscous drag forces on the surface zones of the platform that is part of the time in and part of the time out of the water, we will also find non-zero mean forces. This is illustrated by considering a vertical pile of diameter D in regular deep sea water waves. The horizontal drag force on the pile can be split up into two parts. One part is obtained by integrating forces from $z = -\infty$ to $z = 0$ and the second part comes from integrating forces from $z = 0$ to the instantaneous free-surface elevation ζ. If the free-surface elevation is lower than $z = 0$, the second part corrects the first part being integrated up to $z = 0$, i.e. over too large an area. The mean value of the first part is zero, while

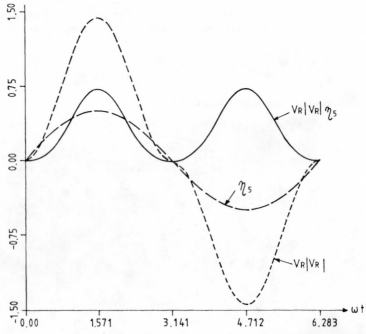

Fig. 5.16. Time variation of expressions in the viscous drift force on a pontoon of a semi-submersible (see equations (5.37) and (5.35)).

the mean value of the second part over one period can be estimated from

$$\frac{\rho}{2}C_D D\overline{\zeta u \, |u|} = \frac{\rho}{2}C_D D\omega^2 \zeta_a^{\,3} \overline{\sin^2 \omega t \, |\sin \omega t|}$$

$$= \frac{2}{3\pi}\rho C_D D\omega^2 \zeta_a^{\,3} \tag{5.38}$$

Depending on the phase angles between the platform motions and the wave motions it is possible that viscous effects create a force that causes the platform to move against the waves. This is not possible according to Maruo's formula based on a potential theory.

Since potential flow effects create a drift-force that is proportional to the square of the wave amplitude and viscous effects cause a drift-force that is proportional to the cube of the wave amplitude, viscous effects become more and more important with increasing wave amplitude. In the following sections we will neglect viscous effects on mean wave loads.

SLOW DRIFT MOTIONS IN IRREGULAR WAVES

Slow-drift motions are resonance oscillations excited by non-linear interaction effects between the waves and the body motion. Due to low damping large motions occur. Fig. 5.17 shows an example of slow drift horizontal motion of a tension leg platform. The first-order wave-induced motion is also present in the recording. It illustrates that slow-drift motions are of equal importance as the linear first-order motions in design of mooring systems for large-volume structures. For a moored structure slow-drift resonance oscillations occur in surge, sway and yaw. For a freely floating structure with low waterplane area second-order slow-drift motions occur also in heave, pitch and roll.

Slow-drift excitation loads are large when the mean wave loads are large. This means slow-drift motions are most important for large volume structures.

A general formula for slow-drift excitation loads can be partly derived in a similar way to the expression for mean wave loads (see equation (5.27)). This means we start out with the results in equation (5.3), generalize this to N wave components and include all second-order contributions. For the slow-drift excitation loads a contribution from the second-order potential is needed. This was not true for mean wave loads. In a later section we will discuss the second-order potential in more detail. At this stage we will concentrate on a general formula for slow-drift excitation loads F_i^{SV} and how this can be further simplified. We can formally write

$$F_i^{SV} = \sum_{j=1}^{N}\sum_{k=1}^{N} A_j A_k [T_{jk}^{\,ic} \cos\{(\omega_k - \omega_j)t + (\epsilon_k - \epsilon_j)\}$$
$$+ T_{jk}^{\,is} \sin\{(\omega_k - \omega_j)t + (\epsilon_k - \epsilon_j)\}] \tag{5.39}$$

where the wave amplitudes A_i, wave frequencies ω_i, random phase angles ϵ_i and number of wave components N are explained by equation (5.25). Further, F_1^{SV}, F_2^{SV} and F_3^{SV} are respectively x-, y-, and z-components of the slow-drift force and F_4^{SV}, F_5^{SV} and F_6^{SV} are moments about the x-, y- and z-axes. The coefficients T_{jk}^{ic} and T_{jk}^{is} can be interpreted as second-order transfer functions for the difference frequency loads. The word 'transfer function' implies that T_{jk}^{ic} and T_{jk}^{is} are independent of the wave amplitudes, but are functions of ω_j and ω_k. They can be calculated independently of the sea state. This is similar to the familiar transfer functions for linear wave-induced motions and loads. There is an ambiguity in defining T_{jk}^{ic} and T_{jk}^{is}, which is easily seen if we consider two wave components and analyse the results in equation (5.3). According to equation (5.39) there is one component with frequency $\omega_2 - \omega_1$ and one component with frequency $\omega_1 - \omega_2$.

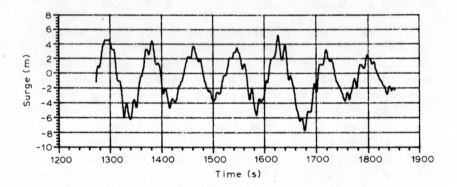

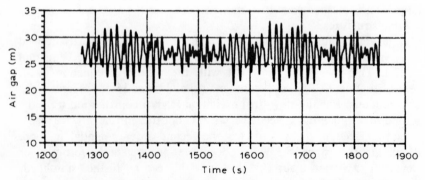

Fig. 5.17. Illustration of horizontal slow-drift surge motion of a tension leg platform. The data are from model tests and show simultaneous recording of air gap (i.e. relative vertical motion between the underside of platform deck and the waves). The latter has a similar dependence as the incident wave field. ($H_{\frac{1}{3}} = 7$ m, $T_0 = 12$ s.)

However, there is only one component with frequency $\omega_2 - \omega_1$ in equation (5.3) which comes out of the analysis. We are free to divide the result into components with frequencies $\omega_1 - \omega_2$ and $\omega_2 - \omega_1$.

In this context we will follow a definition used by Newman (1974), viz.

$$T_{jk}{}^{ic} = T_{kj}{}^{ic}, \qquad T_{jk}{}^{is} = -T_{kj}{}^{is} \qquad (5.40)$$

The mean value of equation (5.39) can be found by noting that the mean value over a long period of any oscillating terms in equation (5.39) is zero. The only time independent terms occur when $k = j$. The result is

$$\overline{F_i^{SV}} = \sum_{j=1}^{N} A_j^2 T_{jj}{}^{ic} \qquad (5.41)$$

i.e. similar to equation (5.27). This means $A_j^2 T_{jj}{}^{ic}$ is the mean wave load in direction i due to incident regular waves of amplitude A_j and circular frequency ω_j.

Figs 5.18 and 5.19 show examples of calculated values of $T_{jk}{}^{ic}$ and $T_{jk}{}^{is}$ as functions of ω_j and ω_k. The functional value of $T_{jk}{}^{ic}$ along the line $\omega_j = \omega_k$ is the same as mean wave load component j divided by wave amplitude squared in regular wave component j (or k).

Newman's approximation
Newman (1974) proposed that $T_{jk}{}^{ic}$ and $T_{jk}{}^{is}$ can be approximated by $T_{jj}{}^{ic}$, $T_{kk}{}^{ic}$, $T_{jj}{}^{is}$ and $T_{kk}{}^{is}$. This reduces the computer time significantly. Another desirable consequence is that we do not need to calculate the second-order velocity potential. The reason why Newman's approximation often gives satisfactory results is that $T_{jk}{}^{ic}$ and $T_{jk}{}^{is}$ generally do not change very much with frequency and that we are interested in the result of $T_{jk}{}^{ic}$ and $T_{jk}{}^{is}$ when ω_j is close to ω_k. A large frequency difference $\omega_j - \omega_k$ gives a smaller oscillation period which is further away from the resonance period of the structure. This means we can approximate $T_{jk}{}^{ic}$ and $T_{jk}{}^{is}$ with its values along the line $\omega_j = \omega_k$. This is obviously less good if a $T_{jk}{}^{ic}$ shows pronounced maxima or minima in the vicinity of the line $\omega_j = \omega_k$. Pronounced maxima occur for instance along the line $\omega_j = \omega_k$ if ω_j is close to the natural frequency of heave motion and the heave damping is low. The reason for this is that the mean wave force $T_{jj}{}^{ic}\zeta_a^2$ depends strongly on the linear wave-induced motions. This was discussed earlier in connection with Fig. 5.5. Newman's approximation implies that

$$T_{jk}{}^{ic} = T_{kj}{}^{ic} = 0.5 \, (T_{jj}{}^{ic} + T_{kk}{}^{ic}) \qquad (5.42)$$

$$T_{jk}{}^{is} = T_{kj}{}^{is} = 0 \qquad (5.43)$$

The direct summation of equation (5.39) is still relatively time consuming. Newman (1974) proposed to approximate equation (5.39) further by approximating the double summation by the square of a single series. This implies that only N terms should be added together at each time step compared to N^2 terms by equation (5.39). The formula can be written as

$$F_i^{\mathrm{SV}} = 2\left(\sum_{j=1}^{N} A_j (T_{jj}^{\mathrm{ic}})^{\frac{1}{2}} \cos(\omega_j t + \epsilon_j)\right)^2 \tag{5.44}$$

This equation includes high-frequency effects that have no physical background. In studying the slow-drift response these terms have no influence. Obviously equation (5.44) requires T_{jj}^{ic} to be positive. A modification can be done if some of the T_{jj}^{ic}-values are zero.

The selection of the number of wave components N requires some

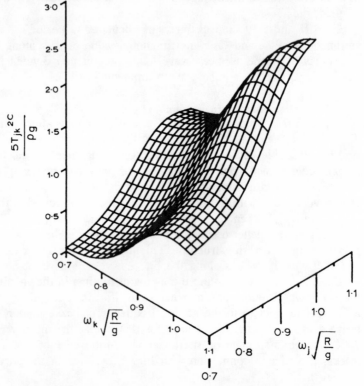

Fig. 5.18. Calculated values of the second-order transfer function T_{jk}^{2c} (see equation (5.39)) for the difference frequency horizontal force on a two-dimensional circular cross-section with an axis in the mean free-surface and radius R. No effect of roll. Infinite water depth. The data are presented as a function of the circular frequencies ω_j and ω_k. (Faltinsen & Zhao, 1989.)

care. We must require that $(\omega_{max} - \omega_{min})/N$ is a small fraction of the natural frequency of the response we want to excite. To avoid the signal repeating itself too soon, it is necessary to choose A_j at random values of ω_j (see discussion following equation (5.25)).

Instead of writing the slow-drift excitation force in a time series, it may be convenient to write it in spectral form. According to Pinkster (1975) we may calculate the spectral density of the low frequency part as

$$S_F(\mu) = 8 \int_0^\infty S(\omega)S(\omega + \mu) \left(\frac{\bar{F}_i\left(\omega + \dfrac{\mu}{2}\right)}{\zeta_a^{\,2}} \right)^2 d\omega \qquad (5.45)$$

where $\bar{F}_i(\omega + \mu/2)$ is the mean wave load in direction i for frequency $\omega + \mu/2$.

Let us now study the response to the slow-drift excitation force and assume that the slow-drift response x can be described by a mass–spring system with linear damping, i.e.

$$m\ddot{x} + b\dot{x} + cx = F_i(t) \qquad (5.46)$$

The slow-drift excitation force $F_i(t)$ is a random time process. The variable x could for instance be the surge motion. We can write the mean square value of x as

$$\sigma_x^{\,2} = \int_0^\infty \frac{S_F(\mu)\,d\mu}{(c - m\mu^2)^2 + b^2\mu^2} \qquad (5.47)$$

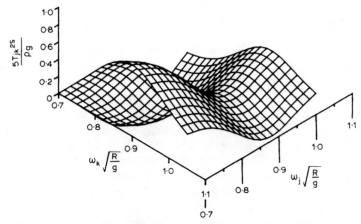

Fig. 5.19. Calculated values of the second-order transfer function T_{jk}^{2s} (see equation (5.39)) for the difference frequency horizontal force on a two-dimensional circular cross-section with an axis in the mean free surface and radius R. No effect of roll. Infinite water depth. The data are presented as a function of the circular frequencies ω_j and ω_k. (Faltinsen & Zhao, 1989.)

The mean square values give a useful measure of the extreme values. Typically σ_x will be $\frac{1}{3}$ to $\frac{1}{4}$ of the extreme values.

Let $\mu_n = \sqrt{(c/m)}$. Then in case of small damping we may approximate equation (5.47) by

$$\sigma_x^2 = S_F(\mu_n) \int_0^\infty \frac{\mathrm{d}\mu}{(c - m\mu^2)^2 + b^2\mu^2}$$

$$= S_F(\mu_n) \frac{\pi}{2cb} \tag{5.48}$$

The reason why we can set $S_F(\mu_n)$ outside the integral sign is that the major contribution to the integral in equation (5.47) comes from the vicinity of $\mu = \mu_n$ when b is small.

This demonstrates that small damping b means large motions and that the magnitude of response depends on how much slow-drift excitation energy exists at $\mu = \mu_n$. By small damping we mean small relative to the critical damping $2m\mu_n$.

We will now use the formula for σ_x to discuss the standard deviation of the slow-drift surge motion of a moored tanker. The slow-drift spectrum S_F for surge can be calculated when we know the longitudinal drift forces in regular waves. The slow-drift spectrum will be wave heading dependent. The drift forces can for instance be calculated by using three-dimensional source technique (see chapter 4) and the equation for conservation of momentum in the fluid (see equation (5.9)). The restoring coefficient c follows by the mooring characteristics and the natural circular frequency by $\mu_n = \sqrt{(c/(M + A_{11}))}$, where M and A_{11} are the mass and the added mass in surge of the ship. The added mass in surge can be calculated by a three-dimensional source technique. It will amount to 10–20% of the mass of the ship. The damping b is due to hull damping and anchor line damping. The hull damping is caused by skin friction, eddy-making and wave-drift damping (Wichers, 1982). Due to the low frequency of oscillation wave radiation damping can be neglected. For higher sea states the wave-drift damping is the dominant hull damping component. As an example, for a 235 m long ship the wave-drift damping amounted to 85% of the total damping for $H_{\frac{1}{3}} = 8.1$ m, while the wave-drift damping was negligible for $H_{\frac{1}{3}} = 2.8$ m. The anchor line may also contribute significantly to the damping (Huse, 1986). This is not accounted for in the present discussion.

For slow-drift oscillations of a TLP or slow-drift oscillations in sway of a ship both eddy-making damping and wave-drift damping have to be included.

Wave-drift damping

Wave-drift damping is caused by the waves and can be seen by comparing free-decay model tests of a ship in still water and in regular waves (see Fig. 5.20). We can try to explain the wave-drift damping in surge by interpreting the slow-drift surge motion as a quasi-steady forward and backward speed. It is well known that the added resistance of a ship in waves is speed dependent. This is for instance evident from

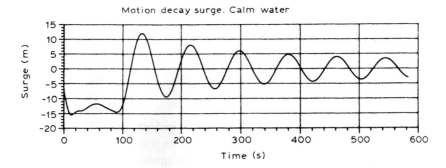

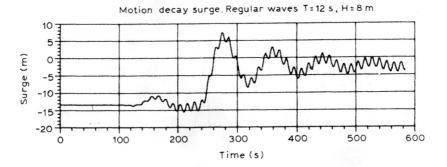

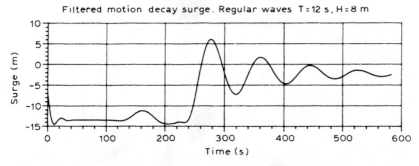

Fig. 5.20. Influence of wave-drift damping demonstrated by free decay tests in surge of a tension leg platform both in calm water and in regular waves. (Wave period $T = 12$ s, wave height $H = 8$ m).

equation (5.22) which is valid in the non-zero speed case when the wavelength relative to the ship length is small. We can now interpret the term that is proportional to the ship speed as a damping term. This means we can write the wave-drift damping theoretically for small wavelengths in head sea as

$$B_{11}{}^{\mathrm{WD}} = \rho \omega_0 \zeta_a{}^2 \int_{L_1} \sin^2 \theta\, n_1\, \mathrm{d}l \qquad (5.49)$$

By using the numerical method outlined by Zhao et al. (1988) it is possible to calculate the wave-drift damping for any wavelength.

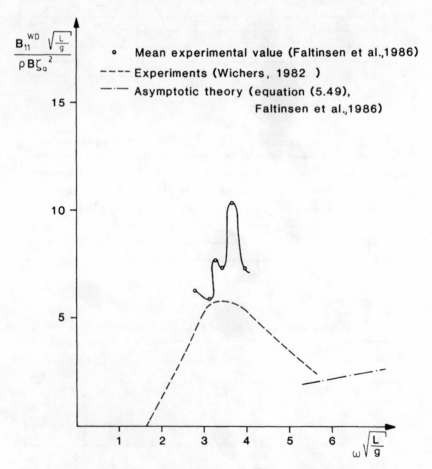

Fig. 5.21. Examples of wave-drift damping $B_{11}{}^{\mathrm{WD}}$ in surge for head sea regular waves incident on a ship. (B = ship beam, L = ship length, ζ_a = wave amplitude of incident waves, ω = circular frequency of oscillations). The results by Faltinsen et al. (1986) and by Wichers (1982) are for different ships.

Experimental results for wave-drift damping in surge for two ship hulls in head sea regular waves are presented in Fig. 5.21. Experimental results can easily show large scatter, which means that high accuracy is needed in the experimental technique. An example of numerical results for a vertical circular cylinder is presented in Fig. 5.22. To obtain the

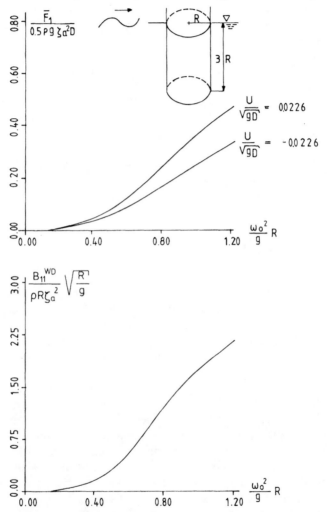

Fig. 5.22. Wave-drift force $\bar{F}_1$ and wave-drift damping B_{11}^{WD} in surge for a vertical circular cylinder that is free to surge in linear motions. Incident regular waves. Infinite water depth. Draught = $3R$. The lower figure is obtained from the results in the upper figure. (ζ_a = wave amplitude of the incident waves, ω_0 = circular frequency of oscillations of incident waves, U = current velocity. Positive current direction is in the wave propagation direction, D = diameter.)

wave-drift damping in an irregular sea we can proceed as for the mean force, i.e. we can use an equation similar to (5.28). From this we find that the wave-drift damping is proportional to $H_{\frac{1}{3}}^2$ in an irregular sea. Going back to the formula for the standard deviation (equation (5.48)) and noting that S_F is proportional to $H_{\frac{1}{3}}^4$ we see that the standard deviation of the slow-drift surge motion is proportional to $H_{\frac{1}{3}}$ for higher sea states. This is a somewhat unexpected result when we talk about a second-order quantity, but experiments show the same tendency.

Faltinsen & Zhao (1989) argued that there also ought to be a slowly-varying wave-drift damping if there is a slowly-varying excitation force. This is a consequence of considering the slow-drift velocity to be quasi-steady. Zhao & Faltinsen (1988) showed that the slowly-varying wave-drift damping had little influence on the standard deviation of the motions while it had a significant effect on the extreme values. The results from the simulations showed that the extreme values of the motions tended to follow a Rayleigh distribution if slowly-varying wave-drift damping was included. This means the most probable largest value x_{max} in a storm of time duration t can be written as

$$x_{max} = \sigma_x \left(2 \log \frac{t}{T_N} \right)^{\frac{1}{2}} \tag{5.50}$$

where T_N is the natural period of the slow-drift response variable. For a storm of duration 10 hours and $T_N = 100\,\text{s}$ this means $x_{max} = 3.43\sigma_x$. However, more work is needed both hydrodynamically and statistically to achieve more accurate and reliable estimates of extreme values for slow-drift motions. Equation (5.50) should be understood as a rough estimate.

To get good estimates of extreme values of slow-drift motions from model tests or numerical simulations a long simulation time is needed. Fig. 5.23 presents results from numerical simulations. (Transient effects have been excluded.) Each record simulates the response of the same system in the same sea state. The differences in the results from one time series to another are due to random selection of phase angles and wave amplitudes in equation (5.25). The standard deviations are nearly the same, while it is obvious that the extreme values differ. About 20 realizations of the same sea state were used to get good estimates of the most probable largest slow-drift motion amplitude.

If eddy-making damping is important and a frequency domain solution is applied (see equation (5.47)), the equivalent linearization technique can be used to approximate the eddy-making damping term. This will be illustrated by studying slow-drift sway motion of a moored ship. For simplicity we will neglect the coupling with yaw and the wave-drift damping in sway. The equation of motion can be written as

$$(M + A_{22})\ddot{y} + B_D\,\dot{y}\,|\dot{y}| + C_{22}y = F_2^{SV} \tag{5.51}$$

Here A_{22} is added mass in sway, C_{22} is restoring coefficient in sway due to the mooring system, F_2^{SV} is the slow-drift excitation force in sway. The slow-drift sway motion is denoted by y. The coefficient B_D can be expressed as $0.5\rho C_D A$ where ρ is the mass density of the water, C_D is a drag coefficient, and A is the frontal area of the submerged structure against the motion. The problem of finding C_D will be addressed in

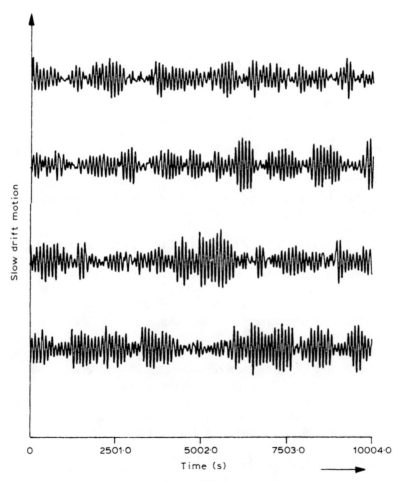

Fig. 5.23. Identical simulations of slow-drift motions of a moored two-dimensional body. The differences in the results are due to random selection of phase angles and wave amplitudes.

chapter 7. We have excluded the relative velocity between the slow-drift motion and the linear relative motion. This could influence the result.

The slow-drift sway is not strictly Gaussian; however, by assuming a Gaussian response, we can write an equivalent damping B^e as (see Price & Bishop, 1974)

$$B^e = B_D \frac{4\sigma_{\dot{y}}}{(2\pi)^{\frac{1}{2}}} \qquad (5.52)$$

We can use equation (5.48) to find that

$$\sigma_{\dot{y}}^2 \approx \mu_n^2 S_F(\mu_n) \frac{\pi}{2C_{22}B^e} = \mu_n^2 S_F(\mu_n) \frac{\pi(2\pi)^{\frac{1}{2}}}{8C_{22}B_D\sigma_{\dot{y}}} \qquad (5.53)$$

This means

$$\sigma_{\dot{y}} = \left(\mu_n^2 S_F(\mu_n) \frac{\pi(2\pi)^{\frac{1}{2}}}{8C_{22}B_D} \right)^{\frac{1}{3}} \qquad (5.54)$$

where $\sigma_{\dot{y}} \approx \mu_n \sigma_y$.

This equation shows that the standard deviation in sway is proportional to $B_D^{-\frac{1}{3}}$, where B_D is like a drag force damping coefficient. This implies low sensitivity between σ_y and B_D. A 100% increase in B_D means only a 20% reduction in σ_y. In equation (5.54) it is only the slow-drift spectrum in sway that depends on the significant wave height. This means the standard deviation of slow-drift sway motion is proportional to $H_{\frac{1}{3}}^{\frac{4}{3}}$.

SLOWLY-VARYING OSCILLATIONS DUE TO WIND

Wind can also produce slowly-varying oscillations of marine structures with high natural periods. This is caused by wind gusts with significant energy at periods of the order of magnitude of a minute. In chapter 2 we discussed how to evaluate the gust spectral density $S(f)$ of horizontal wind speeds (f = frequency in hertz).

In calculating the response to the wind gusts we need first to find an expression for the wind gust force spectrum. We will follow Davenport's (1978) description. We will first consider a structure of frontal area A against the wind, and assume that the structure is sufficiently small so that there is no important variation in the wind over the structure. We can write the horizontal force on the body in the wind direction as

$$F_D = \frac{\rho_{air} C_D}{2} A U^2(t) \qquad (5.55)$$

The mass density of the air ρ_{air} is 1.21 kg m^{-3} at $20\,°C$. Further, $U(t) = \bar{U} + u'$ where $\bar{U}$ is the mean wind velocity and u' is the gust velocity. The mean drag force is

$$\bar{F}_D = \frac{\rho_{air} C_D}{2} A \bar{U}^2 \tag{5.56}$$

Mean wind loads may be of equal importance as the mean wave-drift force. By ignoring terms of order $(u'/\bar{U})^2$ the fluctuating drag forces can be written as

$$F_D'(t) = C_D A \rho_{air} \bar{U} u'(t) \tag{5.57}$$

We note that the fluctuating drag force is *linearly* dependent on the gust velocity. The power spectrum of $F_D'(t)$ is then related to the gust velocity spectrum by

$$S_F(f) = (C_D A \rho_{air} \bar{U})^2 S(f) \tag{5.58}$$

If the size of the structure is not small from a wind field point of view, Davenport introduces an 'aerodynamic admittance function' as a correction of the result. (For more details see Davenport, 1978).

In analysing the response of the structure we can now proceed as we did for the slow-drift oscillations caused by waves. For instance, if we consider head wind the mean square value of the surge motion is

$$\sigma_x^2 = S_F^W(\omega_n) \frac{\pi}{2cb} \tag{5.59}$$

The index W means wind. The relation between gust spectrum expressed respectively by circular frequency ω and frequency f in hertz is

$$S_F^W(\omega)\, d\omega = S_F(f)\, df, \qquad \omega = 2\pi f.$$

This means

$$S_F^W(\omega) = \frac{1}{2\pi} S_F(f) \tag{5.60}$$

The relation between the standard deviation σ_x of the slowly varying oscillations due to wind gust and the mean offset $\bar{x}$ due to steady wind can be expressed in a simple way by equations (5.59) and (5.56). By using the Harris wind spectrum (see equation (2.40)), we find

$$\frac{\sigma_x}{\bar{x}} = \frac{(0.038 \tilde{f}_N)^{\frac{1}{2}}}{(2 + \tilde{f}_N^2)^{\frac{5}{6}}} \cdot \frac{1}{\sqrt{p}} \tag{5.61}$$

Here $\tilde{f}_N = f_N 1800/\bar{U}_{10}$ (with dimensions given in meters and seconds), $\bar{U}_{10} = \bar{U}$ and p is the fraction between the damping b and critical damping $2m\omega_N$ (see equation (5.48)). If for instance $\bar{U}_{10} = 40$ m s^{-1},

$T_N = 100$ s and the damping b is 10% of the critical damping it means that σ_x is 30% of the mean offset $\bar{x}$.

SUM-FREQUENCY EFFECTS

It follows from the discussion of equation (5.3) that there are non-linear effects due to the quadratic velocity term in Bernoulli's equation that can create excitation forces with higher frequencies than the dominant frequency components in a wave spectrum. This is due to terms oscillating with frequencies $2\omega_1$, $2\omega_2$ and $(\omega_1 + \omega_2)$. These may be important for exciting the resonance oscillations in the heave, pitch and roll of a tension leg platform. However, it can be shown that the contribution from the quadratic velocity term in Bernoulli's equation to the sum-frequency heave forces for a tension leg platform is normally small at the natural heave period in heave. This follows from the exponential depth decay of this term, together with the fact that it is the pontoons and the bottom of the columns that normally contribute to the vertical force on a TLP and that the natural periods are from 2 to 4 s. The most important contribution to the vertical sum-frequency force on a TLP comes from the second-harmonic part of the second-order potential (Kim & Yue, 1988). The second-order potential ϕ_2 follows from solving a boundary value problem with the inhomogeneous free-surface condition

$$\phi_{2tt} + g\phi_{2z} = -\frac{\partial}{\partial t}(\phi_{1x}{}^2 + \phi_{1y}{}^2 + \phi_{1z}{}^2)$$

$$+ \frac{1}{g}\phi_{1t}\frac{\partial}{\partial z}(\phi_{1tt} + g\phi_{1z}) \quad \text{on} \quad z = 0 \qquad (5.62)$$

where the index 1 denotes first-order potential. The effect of current is neglected in equation (5.62). Equation (5.62) expresses the fact that the first-order potential ϕ_1 gives an effect on ϕ_2 which can be interpreted physically as an imposed pressure on the free-surface. The existence of second-harmonic pressure at large depth has been explained by Newman (1990). A similar phenomenon exists for the second-order oscillatory wave field when there are two linear long-crested wave fields propagating in opposite directions. We will show this by writing the linear (first-order) potential as

$$\phi_1 = \frac{ga_1}{\omega_1}e^{k_1z}\cos(\omega_1 t - k_1 x + \delta_1)$$

$$+ \frac{ga_2}{\omega_2}e^{k_2z}\cos(\omega_2 t - Ak_2 x + \delta_2) \qquad (5.63)$$

where $A = \pm 1$ depending on the propagation direction of the second wave in equation (5.63). A particular solution that satisfies equation

(5.62), the deep water conditions and the Laplace equation can be written

$$\phi_2 = \frac{2a_1a_2\omega_1\omega_2(\omega_1 - \omega_2)}{-(\omega_1 - \omega_2)^2 + g\,|k_1 - k_2|}\,e^{|k_1 - k_2|z}\,\sin[(\omega_1 - \omega_2)t$$

$$-(k_1 - k_2)x + \delta_1 - \delta_2] \quad \text{for} \quad A = 1 \qquad (5.64)$$

and:

$$\phi_2 = \frac{2a_1a_2\omega_1\omega_2(\omega_1 + \omega_2)}{-(\omega_1 + \omega_2)^2 + g\,|k_1 - k_2|}\,e^{|k_1 - k_2|z}\,\sin[(\omega_1 + \omega_2)t$$

$$-(k_1 - k_2)x + \delta_1 + \delta_2] \quad \text{for} \quad A = -1 \qquad (5.65)$$

Equations (5.64) and (5.65) show that it is only when the waves propagate in opposite directions that there are sum-frequency effects.

If the two frequencies ω_1 and ω_2 are equal, we note that the second-harmonic term does not vary with x and z. This interesting result was discussed by Longuet-Higgins (1953). When ω_1 is close to ω_2, we note that the sum-frequency effects decay slowly with depth. The consequence of the small depth decay is that the resulting fluid velocities are negligible. It is the pressure that is of importance. However, since resulting pressure gradients are small, there are small forces on a completely submerged body.

In the case of regular waves incident on a two-dimensional body, the reflected waves will interact with the incident waves in the way represented by equation (5.65). In this case $\omega_1 = \omega_2$. For the transmitted wave system, the linear waves generated by the body propagate in the same direction as the incident waves and there is no effect from the second-order potential in the far-field. For a three-dimensional flow problem the waves generated by the body along a ray opposite to the incident wave direction create a similar second-order depth effect to that in the two-dimensional case with the reflected wave system. Since the interaction along the ray opposite to the incident wave direction is affected by interaction along other oblique rays, the second-order effects are not equal in the two-dimensional and the three-dimensional cases. By assuming that the dominant contribution to the second-order velocity potential is from the far-field part of the forcing function in the free-surface boundary condition, Newman (1990) showed that the second-order harmonic potential is inversely proportional to the depth and not exponentially decaying for large depth. This is in agreement with Kim & Yue's numerical results. In the case of two linear frequency components ω_1 and ω_2, the $2\omega_1$- and $2\omega_2$-terms decay inversely proportional to the depth. The $(\omega_1 + \omega_2)$-term decays inversely proportional to the depth if the two frequency components are close to each other.

Table 5.1. *Main characteristics of a TLP presented in Fig.* 3.15

Water depth, m	600
Displacement, tonnes	52.000
Distribution, tonnes	
Deck structures	6.000
Facilities	10.000
Ballast	1.000
Hull	16.000
⟨Subtotal⟩	⟨33.000⟩
Top tension	
Σ Risers	2.000
Σ Tethers	17.000
Dimensions, m	
Length overall	87
Column spacing	70
Draught	32
Under deck clearance	25
Column diameter	17
Natural period, s	
Sway, surge	107
Heave	2.2
Pitch, roll	2.4
Yaw	87

EXERCISES

5.1 Tension leg platform
A tension leg platform is shown in Fig. 3.15. Main characteristics are given in Table 5.1.

(a) Show that the natural periods of surge and heave motion of a tension leg platform can be approximated as

$$T_{n1} = 2\pi \left(\frac{M + A_{11}}{P_0/l} \right)^{\frac{1}{2}}$$

$$T_{n3} = 2\pi \left(\frac{M + A_{33}}{EA/l} \right)^{\frac{1}{2}}$$

where P_0 is the pre-tension in the tethers, l the length of the tethers and A is the cross-sectional dimensions of the tethers.

(b) Consider incident regular waves propagating in the positive x-direction. The mean wave-drift forces in surge are given in Fig. 5.24 as a function of ω. Viscous effects are excluded.

Discuss the behaviour of the wave-drift forces, particularly asymptotic values and cancellation effects. Compare with the asymptotic formulas for small wavelengths.

(c) Discuss in a qualitative way the importance of viscous effects on the mean wave-drift forces.

(d) Calculate the standard deviation of the slow-drift surge motion due to wind when the wind speed is 40 m s^{-1}. Make your own choices when necessary data are not given. Give an explanation of your choices. You may assume that the damping is due to drag forces.

(e) Discuss in a qualitative sense the importance of wave-drift damping on slow-drift surge motion due to wind.

5.2 Moored ship

Consider a moored ship of length L in long-crested head sea waves that can be described by a wave spectrum $S(\omega)$. In the following calculations use

$$\frac{S(\omega)}{H_{\frac{1}{3}}^{2} T_1} = \begin{cases} (32\pi)^{-1} & 0.5 < \dfrac{\omega T_1}{2\pi} < 1.5 \\ 0 & \text{for other frequencies} \end{cases}$$

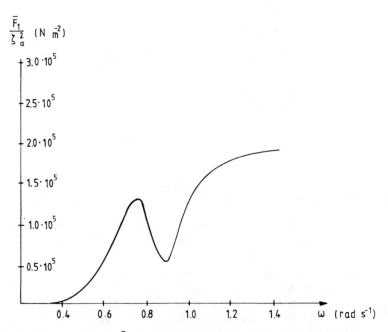

Fig. 5.24. Wave-drift force $\bar{F}_1$ of the TLP described in Fig. 3.15 and Table 5.1. 0° wave heading. (ω = circular frequency of oscillations, ζ_a = wave amplitude of the incident waves.)

Assume the longitudinal wave drift force $\bar{F}_1$ in regular head sea waves of wavelength λ and wave amplitude ζ_a can be written in the simplistic form

$$\frac{\bar{F}_1}{\rho g L \zeta_a^2} = \begin{bmatrix} 0.03 & 0 < \lambda/L < 0.5 \\ 0 & \text{for other wavelengths} \end{bmatrix}$$

(a) Show that the mean longitudinal wave force in an irregular sea can be written as

$$\bar{F}_1 = 0.0006 \rho g L H_{\frac{1}{3}}^2 T_1 (\omega_{max} - \omega_{min})$$

where $\omega_{max} = 3\pi/T_1$ and ω_{min} the largest value of $2\sqrt{\pi g/L}$ and π/T_1 whenever those values are less than ω_{max}. Otherwise $\omega_{min} = \omega_{max}$.

(b) Use equation (5.48) to calculate the standard deviation σ_x of the slow-drift surge motion. Show that the slow-drift excitation spectrum $S_F(\mu_n)$ can be expressed as

$$S_F(\mu_n) = 2(0.0006 \rho g L H_{\frac{1}{3}}^2 T_1)^2 (\omega_2 - \omega_1)$$

and explain the meaning of ω_1 and ω_2. Assume the slow-drift damping is only due to wave-drift damping. Express the slow-drift damping coefficient B_{11}^{WD} in regular waves in the following simplistic way

$$\frac{B_{11}^{WD}}{\rho(Lg)^{\frac{1}{2}}\zeta_a^2} = \begin{bmatrix} 0.7 & 0 < \lambda/L < 0.5 \\ 0 & \text{for other wavelengths} \end{bmatrix}$$

Show that the slow-drift damping $\overline{B_{11}^{WD}}$ in irregular seas can be written as

$$\overline{B_{11}^{WD}} = 0.014 \rho (Lg)^{\frac{1}{2}} H_{\frac{1}{3}}^2 T_1 (\omega_{max} - \omega_{min})$$

Show that

$$\sigma_x = 0.0014 T_n H_{\frac{1}{3}} \left[\frac{\rho T_1 (Lg)^{\frac{3}{2}} (\omega_2 - \omega_1)}{(M + A_{11})(\omega_{max} - \omega_{min})} \right]^{\frac{1}{2}}$$

where $T_n = 2\pi/\mu_n$ and M and A_{11} are respectively the mass of the ship and the added mass in surge of the ship.

5.3 Added resistance of ships in waves

(a) Equation (5.24) is a crude approximation for the added resistance of a ship in the 'ship motion' range. Use this formula to discuss how the added resistance depends on main hull characteristics. (Hint: See relevant discussion of heave and pitch motion in chapter 3.)

(b) The added resistance R_{AW} of a ship is normally non-dimensionalized as $R_{AW}/(\rho g \zeta_a^2 B^2/L)$. Another way of non-dimensionalizing R_{AW} is to divide R_{AW} by $\rho g \zeta_a^2 B$. Which of the two non-dimensionalized values of added resistance will show least sensitivity to main hull characteristics?

(c) Equation (5.28) can be used to calculate the added resistance of a ship in a short-term sea state where the waves are long-crested. Consider situations where the 'ship motion' range of the added resistance contributes most to the integral in equation (5.28). Approximate the integral as

$$\overline{F_1^s} = 2S(\omega_n) \int_0^\infty \left(\frac{\bar{F}_1(\omega; \beta)}{\zeta_a^2} \right) d\omega \qquad (5.66)$$

where $\bar{F}_1(\omega, \beta)$ only contains the ship-motion part of the added resistance. This means the bow-wave reflection part of added resistance for small wavelengths is excluded. In this way the integral in equation (5.66) converges. ω_n in equation (5.66) is the frequency where the non-dimensionalized added resistance in regular waves has a peak value (see Fig. 5.11). Assume a modified Pierson–Moskowitz spectrum with two parameters $H_{\frac{1}{3}}$ and T_1 (see equation (2.24)) and head sea waves. Select a value of ω_n. Discuss for what ship length and ship speed your choice of ω_n is realistic. Use Table 2.2 for the joint frequency of significant wave height and spectral peak period representative for the northern North Sea. Assume the integral in equation (5.66) is known. Discuss how much the added resistance for a ship can vary for realistic combinations of $H_{\frac{1}{3}}$ and T_1.

(d) Consider an infinitely long horizontal cylinder of arbitrary cross-section in beam sea regular waves. Assume the cylinder has no mean velocity and the current velocity is zero. Use Gerritsma & Beukelman's formula (equation (5.23)) and show that it can be written in the same form as Maruo's formula (equation (5.16)). (Hint: Use equation (3.26).) What effects are included in Maruo's formula, but not in Gerritsma & Beukelman's formula?

6   CURRENT AND WIND LOADS

Analyses of current and mean wind loads have many similarities even though water and air have different physical properties and typical current and wind velocities differ significantly in magnitude (see Table 6.1).

The main factors influencing the flow can be characterized in terms of the *Reynolds number* $\mathrm{Rn} = UD/v$ (U = characteristic free-stream velocity, D = characteristic length of the body, v = kinematic viscosity coefficient), and the *roughness number* $= k/D$ (k = characteristic dimension of the roughness on the body surface). Also important are *body form, free-surface effects, sea floor effects, nature and direction of ambient flow relative to the structure's orientation,* and *reduced velocity* ($=U/(f_n D)$) for an elastically mounted cylinder with natural frequency f_n). With nature of ambient flow we mean, for instance, that the turbulence level in the incident flow can affect the forces on the body.

All full-scale cases of practical interest for our purpose are high Reynolds number flow. For instance, the Reynolds number for current flow with velocity $1 \mathrm{~m~s}^{-1}$ past a column of a semi-submersible with diameter 15 m is $1.4 \cdot 10^7$ at 20 °C. The Reynolds number for wind flow with velocity $40 \mathrm{~m~s}^{-1}$ past a structural part with characteristic length 20 m is $5.3 \cdot 10^7$ at 20 °C.

It is difficult to predict theoretically current and wind loads with sufficient accuracy. This is often true for three-dimensional flow and in turbulent incoming flow. The latter occurs for instance for pipelines close to the sea floor and for structures in the wake of other structures like jackets and flare towers.

In the following text we will summarize the important physical characteristics of separated flow around structures in incident steady flow. In chapter 7 we will discuss the effect of unsteady ambient flow. We begin by examining steady incident flow past a circular cylinder.

STEADY INCIDENT FLOW PAST A CIRCULAR CYLINDER

A classic problem in non-separated potential theory is steady flow past a fixed circular cylinder. The tangential velocity U_e around the cylinder

Table 6.1. *Mass density (ρ) and kinematic viscosity (v) of water and air*

Temperature	Fresh water ρ (kg m^{-3})	Fresh water $v \cdot 10^6$ (m^2 s^{-1})	Salt water (salinity 3.5%) ρ (kg m^{-3})	Salt water (salinity 3.5%) $v \cdot 10^6$ (m^2 s^{-1})	Dry air ρ (kg m^{-3})	Dry air $v \cdot 10^6$ (m^2 s^{-1})
0 °C	999.8	1.79	1028.0	1.83	1.29	13.2
5 °C	1000.0	1.52	1027.6	1.56	1.27	13.6
10 °C	999.7	1.31	1026.9	1.35	1.25	14.1
15 °C	999.1	1.14	1025.9	1.19	1.23	14.5
20 °C	998.2	1.00	1024.7	1.05	1.21	15.0

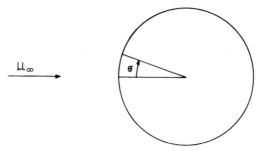

Fig. 6.1. Definition of steady incident flow past a circular cylinder.

surface is then

$$U_e = 2U_\infty \sin \theta \tag{6.1}$$

where U_∞ is the incident flow velocity and θ is an angular coordinate with $\theta = 0$ corresponding to the forward stagnation point on the cylinder surface (see Fig. 6.1). The pressure p on the body follows from Bernoulli's equation, i.e.

$$p + \tfrac{1}{2}\rho U_e^2 = p_0 + \frac{\rho}{2} U_\infty^2 \tag{6.2}$$

where p_0 is the ambient pressure. The resulting pressure coefficient C_p can be written as

$$C_p \equiv \frac{p - p_0}{\tfrac{1}{2}\rho U_\infty^2} = 1 - 4 \sin^2 \theta \tag{6.3}$$

This is shown in Fig. 6.2, where experimental data for three different Reynolds numbers are also plotted. In reality the problem is unsteady. This means the experimental pressure values in Fig. 6.2 are time averaged data. The data are only presented for θ-values between 0 and

180°. However, the pressure values for a negative θ-value between 0 and $-180°$ are symmetric about $\theta = 0$. This means there is no mean force on the cylinder in the lift direction, i.e. orthogonal to the current direction.

It is evident from Fig. 6.2 that non-separated potential theory is a poor approximation to the real flow. The figure shows that the pressure coefficient and therefore also the drag coefficient C_D is strongly Reynolds number dependent. We note that the experimental pressure is nearly constant over a major part of the lee side of the cylinder. We can use this observation to express the force on the cylinder in terms of a 'base underpressure' coefficient

$$Q = \frac{p_B - p_0}{\frac{1}{2}\rho U_\infty^2} \qquad (6.4)$$

where p_B is a constant 'base pressure'. This can be done approximately by assuming potential flow on the upstream side of the cylinder, i.e. for θ between $-\pi/2$ and $\pi/2$. For all other θ-values we assume a constant pressure p_B. This is obviously in error for θ-values in the vicinity of

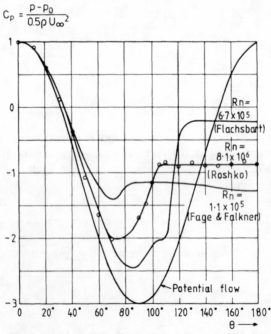

Fig. 6.2. Average pressure distribution p around a smooth circular cylinder in steady incident flow. ($Rn = U_\infty D/\nu$, D = cylinder diameter, U_∞ = incident flow velocity, θ defined in Fig. 6.1, p_0 = ambient pressure.) (Roshko, 1961.)

$\pm\pi/2$. However, the major contribution to the drag force comes from integrating pressure forces in the vicinity of $\theta = 0$ and $|\theta| = \pi$. Further shear forces can be neglected relative to pressure forces. Achenbach (1968) showed that shear forces on a circular cylinder were between 1 and 3% of the pressure forces. The drag force F_1, i.e. the force in the incident flow direction, can be written approximately as

$$F_1 = \tfrac{1}{2}\rho U_\infty^2 \int_{-\pi/2}^{\pi/2} (1 - 4\sin^2\theta)\cos\theta\, R\, d\theta$$

$$+ \int_{\pi/2}^{3\pi/2} (p_B - p_0)\cos\theta\, R\, d\theta \qquad (6.5)$$

This means the drag coefficient C_D can be written as

$$C_D \equiv \frac{F_1}{\tfrac{1}{2}\rho U_\infty^2 D} = (-Q - \tfrac{1}{3}) \qquad (6.6)$$

From Fig. 6.2 we note that $Q = -1.25$ when $\mathrm{Rn} = 1.1 \cdot 10^5$. From equation (6.6) it means $C_D = 0.92$. This is in qualitative agreement with accepted experimental values. If we had used potential theory also on the downstream side of the cylinder, the total force would be zero (D'Alemberts paradox).

Fig. 6.3 shows experimental values for C_D over a broad range of Reynolds number. The effect of surface roughness is included. In practice surface roughness due to marine growth in the North Sea may represent a thickness of 10 cm down to 40 m below mean water level. For larger depths a thickness of 5 cm is a more representative value.

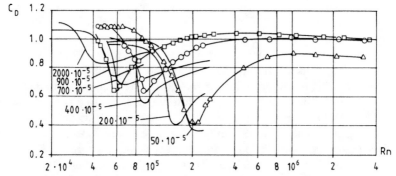

Fig. 6.3. Drag coefficient C_D of rough circular cylinders in steady incident flow for different surface roughness values k/D (k = average height of surface roughness, D = cylinder diameter, $\mathrm{Rn} = U_\infty D/v$, U_∞ = incident flow velocity.) $\triangle$, $k/D = 110 \cdot 10^{-5}$; $\bigcirc$, $k/D = 450 \cdot 10^{-5}$; $\square$, $k/D = 900 \cdot 10^{-5}$; ——, Fage & Warsap (1929), (Achenbach, 1971).

In the figure we note a very distinct drop in C_D-values in a certain Reynolds number range. This is referred to as the critical flow regime and is particularly marked for flow around a smooth cylinder. It is common practice to divide this dependence on Reynolds number into different flow regimes. However, there are many different definitions in the literature. We will refer to the following four different flow regimes: subcritical flow, critical flow, supercritical flow and transcritical flow. For flow around a smooth circular cylinder the subcritical flow regime is for Reynolds number less then $\approx 2 \cdot 10^5$. The critical flow regime is for $\approx 2 \cdot 10^5 < \mathrm{Rn} < \approx 5 \cdot 10^5$. The supercritical flow regime is for $\approx 5 \cdot 10^5 < \mathrm{Rn} < \approx 3 \cdot 10^6$ and the transcritical flow is for Reynolds numbers larger than $3 \cdot 10^6$. In the subcritical flow the boundary layer is always laminar, while in supercritical and transcritical flow the boundary layer is turbulent upstream of the separation point.

BOUNDARY LAYERS

We will give a short description of boundary layer flow and flow separation.

According to the inviscid theory for an ideal fluid we can write the tangential velocity around the cylinder surface in the form of equation (6.1). For θ-values between $-\pi/4$ and $\pi/4$ and for all Reynolds numbers of practical interest this is a good approximation at a small distance from the cylinder surface. However, it cannot be valid in the immediate vicinity of the body surface. The physical tangential velocity has to be zero on the cylinder surface. This means there is a boundary layer along the cylinder surface. Viscous forces are important in this part of the fluid. Let us study the boundary layer in more detail and assume the boundary layer to be laminar, i.e. 'well organized'. As an exercise we will first say that the flow does not separate. That means we will assume equation (6.1) to be valid outside the boundary layer. If we now solve the boundary layer equations (see for instance Schlichting, 1979) we will find a tangential velocity distribution within the boundary layer as shown in Fig. 6.4. This figure gives information on how thick the boundary layer is. However, there are different definitions of boundary layer thickness. One definition is the normal distance from the wall to the point where the tangential velocity is 99% of the local free stream velocity. If potential flow theory is valid, the local free stream velocity is given by equation (6.1). From Fig. 6.4 we see that for $\theta = 80°$ the tangential velocity is 99% of the potential flow velocity when

$$\frac{y}{R}\left(\frac{U_\infty R}{\nu}\right)^{\frac{1}{2}} \approx 2 \tag{6.7}$$

Here y is a coordinate normal to the body surface with $y = 0$ at the body.

For $Rn = 10^5$ this means that the boundary layer thickness is about 1% of the radius.

From the tangential velocity distribution in the boundary layer we can determine where the flow separates. Flow separation means that the flow breaks strongly away from the body. It occurs at a point S on the cylinder surface where there is backflow in the boundary layer on the downstream side of S and no backflow on the upstream side of S. This means

$$\frac{\partial u}{\partial y} = 0 \qquad (6.8)$$

on the body surface at the separation point. If laminar boundary layer flow is assumed and the velocity distribution given by equation (6.1) is valid outside the boundary layer, it follows that the flow around a circular cylinder will separate at angles approximately equal to $\pm110°$ (Schlichting, 1979). However, this is not what happens in reality for

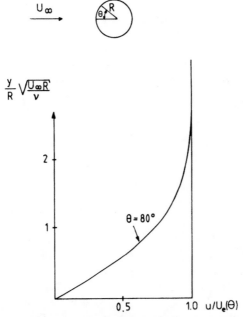

Fig. 6.4. Example of tangential velocity distribution u inside a steady laminar boundary layer flow around a circular cylinder. The results are for one angular position θ and are presented as a function of the y-coordinate normal to the body surface ($y = 0$ is at the body surface). $U_e(\theta)$ = tangential potential flow velocity just outside the boundary layer at the same angular position θ.

laminar boundary layer flow past a circular cylinder. The correct separation angles are approximately $\pm 80°$. This shows that equation (6.1) derived from non-separated potential theory does not represent the velocity field and pressure field outside the boundary layer along the whole cylinder surface. This was also evident from Fig. 6.2. If we had used the experimental pressure distribution in the boundary layer calculations we would have ended up with the correct separation points.

The boundary layer flow will only be laminar up to a certain Reynolds number, beyond which transition to turbulence occurs. In order to illustrate this let us define a local Reynolds number $Rn_x = U_\infty x / v$, where x is the distance along the cylinder surface from the forward stagnation point. At a certain value of $Rn_x = Rn_{crit}$ the flow will become unstable. Downstream of this point small disturbances will be amplified. Fig. 6.5 shows where the point of instability occurs for flow past a circular cylinder. This is based on calculations. However, we should note that the flow outside the boundary layer has been represented by equation (6.1). This is the reason why the separation point S is not correct. The figure tells us that the point of instability for Reynolds number 10^4 is slightly ahead of the separation point. We note that with increasing Reynolds number the point of instability occurs further upstream on the cylinder. However, the flow will not become turbulent immediately downstream of the instability point. There is a certain transition range before the flow becomes turbulent. The distance from the point of instability to the point where the turbulent flow starts has to be experimentally determined.

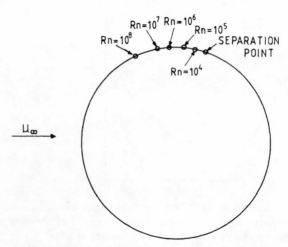

Fig. 6.5. Position of points of instability against the Reynolds number Rn for steady incident flow past a circular cylinder. (Adapted from Schlichting, 1979.)

The separation points for turbulent boundary layer flow differs from laminar boundary layer. For supercritical and transcritical flow where the boundary layer is always turbulent upstream of the separation points, the separation angles are about ±120°. From this one sees that the flow separates more easily in laminar boundary layer flow than in turbulent layer flow. This is a characteristic for any body shape.

WAKE BEHAVIOUR

Up to now we have concentrated on what is happening on the cylinder surface. Only the time-averaged flow has been discussed. In reality the flow is unsteady and eddies are alternatively shed from each side of the cylinder and convected with the flow. The vorticity of the eddies is created in the boundary layer of the body.

Fage & Johansen (1928) studied experimentally the vorticity distribution in the separated flow close to the separation points. The vorticity was found to be concentrated in thin sheets leaving the separation points. At some distance behind the body, the sheets were observed to break up into trails of discrete vortices. Most details were presented for cross-flow past a flat plate. The vorticity was estimated by measuring the flow velocity and estimating the vorticity from the two-dimensional definition of vorticity, i.e. by

$$\omega_z = \frac{\partial v}{\partial x} - \frac{\partial u}{\partial y} \tag{6.9}$$

(see equation (2.2)). The area of vorticity close to the flat plate and outside the boundary layer is presented in Fig. 6.6. An example of the velocity and vorticity distribution across the vortex sheet, i.e. the area of

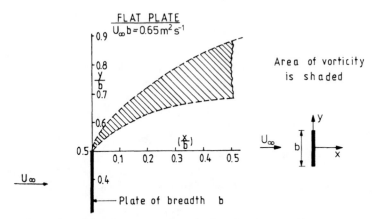

Fig. 6.6. Area of vorticity close to an edge of a flat plate in cross-flow. (Adapted from Fage & Johansen, 1928.)

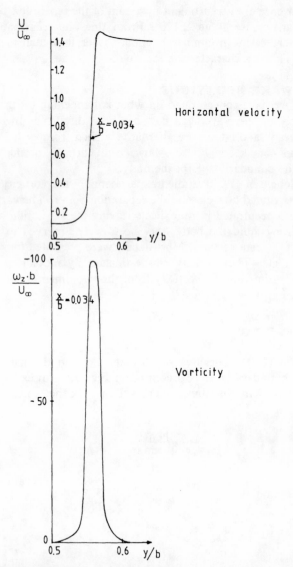

Fig. 6.7. Example of horizontal velocity (u) and vorticity (ω_z) distribution across the area of vorticity in Fig. 6.6. ω_z is defined by equation (6.9). Notation defined in Fig. 6.6. (Adapted from Fage & Johansen, 1928.)

vorticity, is presented in Fig. 6.7. It is evident that the velocity has a strong gradient across the vortex sheet and that the flow velocity is small on the back side of the body.

The amount of vorticity passing a section of the sheet per unit time can be estimated by

$$K = \int_{y_2}^{y_1} \left(\frac{\partial v}{\partial x} - \frac{\partial u}{\partial y} \right) u \, dy \tag{6.10}$$

where y_1 and y_2 are y-coordinates outside the vortex sheet on the upper and lower side, respectively.

When the vortex sheet is thin, we can neglect $\partial v/\partial x$ relative to $\partial u/\partial y$. This means equation (6.10) can be written

$$K = -\int_{y_2}^{y_1} \frac{\partial u}{\partial y} u \, dy = -\tfrac{1}{2}(U_1^2 - U_2^2) \tag{6.11}$$

Here U_1 and U_2 are respectively velocities at the outer and inner boundaries of the vortex sheet. Equation (6.11) applies also in the boundary layer. In this case y means the coordinate normal to the cylinder surface and u is the velocity component tangential to the cylinder surface. If equation (6.11) is applied at the separation point, we can write it as

$$K = -\tfrac{1}{2}U_s^2 \tag{6.12}$$

where U_s is the tangential velocity just outside the boundary layer on the local upstream side.

If we imagine the vorticity stays concentrated in a vortex sheet leaving each separation point, we can relate $\tfrac{1}{2}U_s^2$ to the rate of change of circulation $\partial \Gamma/\partial t$ around the vortex sheet. The circulation Γ can be written as

$$\Gamma = \oint_C \mathbf{V} \cdot d\mathbf{s} \tag{6.13}$$

where $\mathbf{V}$ is the fluid velocity and $d\mathbf{s}$ is an element of C. The integration curve C has to be closed, intersect the separation point and otherwise be outside the analysed vortex sheet and not intersect other vortex sheets. This is illustrated in Fig. 6.8, where we for simplicity have assumed that the vortex sheet (free-shear layer) is infinitely thin. The integration direction in equation (6.13) is in the counter-clockwise direction. Then for the vortex sheets leaving the two separation points A and B in Fig.

6.8 we can write

$$\frac{\partial \Gamma_a}{\partial t} = -\tfrac{1}{2} U_{sa}^{2} \tag{6.14}$$

$$\frac{\partial \Gamma_b}{\partial t} = \tfrac{1}{2} U_{sb}^{2} \tag{6.15}$$

The circulation in a sheet leaving points A and B follows from an integration of equations (6.14) and (6.15), respectively, with respect to time. In a flow which starts from rest and with no initial vorticity in the flow both Γ_a and Γ_b are initially zero.

It should be noted that the circulation around a curve that circumscribes the cylinder and does not intersect the vortex sheets, has to be zero according to Kelvin's theorem (Newman, 1977).

Free-shear layers easily become unstable and turbulent. If the flow is laminar at the separation point, it will take only a short distance from the separation point before it becomes unstable and turbulent. This means wakes are turbulent even if the boundary layer flow is laminar.

VORTEX SHEDDING

In the starting process of separated flow around a circular cylinder a symmetric wake picture develops, but due to instabilities asymmetry will soon occur. The consequence is that vortices are alternatively shed from each side of the cylinder.

Von Karman (see Lamb, 1945: pp. 224–9) has studied the stability of

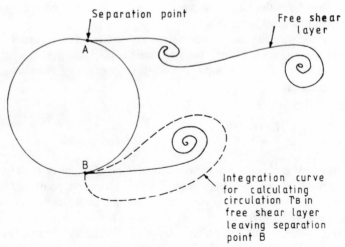

Fig. 6.8. Example of a curve used to calculate the circulation Γ of a free shear layer leaving a separation point.

idealized vortex streets. He investigated the stability of certain geometrical configurations of the eddies created behind a two-dimensional bluff body. Each eddy was mathematically represented as a point vortex of strength magnitude $|\Gamma|$. The eddies were situated in two parallel rows. The directions of rotation of the eddies in one row are uniform and opposite to the rotation in the other row. Von Karman found that only two different arrangements are possible. The eddies in one row are either placed exactly opposite those of the other row or they are symmetrically staggered (Fig. 6.9). The stability investigation lead to the result that the first arrangement is unstable. The second arrangement is generally unstable, but become stable for a very definite ratio between the vortex street width h and distance l between two adjacent vortices in the same row, i.e.

$$\frac{h}{l} = \frac{1}{\pi} \cosh^{-1} \sqrt{2} = 0.28 \tag{6.16}$$

If we use a coordinate system fixed to the body, the vortices will move with a velocity

$$U_\infty - \frac{\Gamma}{l\sqrt{8}} \tag{6.17}$$

Here U_∞ is the incident velocity far upstream.

We can use von Karman's idealized vortex street to get an estimate of the vortex shedding period T_v. From Fig. 6.9 we find

$$T_v\left(U_\infty - \frac{\Gamma}{l\sqrt{8}}\right) = l \tag{6.18}$$

It should be noted Γ and l cannot be estimated from von Karman's theory. The reason is that there is no body geometry involved. Von Karman's theory is not limited to a circular cylinder. For simplicity let us set $h = D$ where D is the cylinder diameter and let us approximate the vortex velocities by U_∞. We may write

$$T_v U_\infty = \frac{D}{0.28} \tag{6.19}$$

The non-dimensional vortex shedding frequency may be represented by

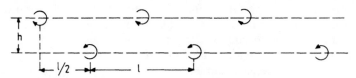

Fig. 6.9. Arrangement of vortices in a von Karman vortex street.

the Strouhal number (St) and defined by

$$St = \frac{f_v D}{U_\infty} \tag{6.20}$$

where $f_v = 1/T_v$. We then find $St = 0.28$. This is a reasonable value for transcritical flow. However it does not agree with experimental values for subcritical flow. Better agreement is found if we use $h \approx 1.2D$ and $U_\infty - \Gamma/(l\sqrt{8}) \approx 0.85U_\infty$. This leads to $St = 0.2$. This is a reasonable value for subcritical flow.

We should note that we do not always see a vortex street behind the body neither is there always a single vortex shedding frequency. In critical and supercritical flow there is a spectrum of vortex shedding frequencies.

In subcritical flow past a circular cylinder we may divide the wake into three parts. The first is the so-called formation region $(l_{fr} < \approx 5R)$; the second, the stable region where the vortices exhibit the characteristics of a fairly uniform vortex street $(5R \leq l_{sr} \leq 12R)$ and the third, the unstable region extending beyond $12R$. The lateral spacing of the vortex cores gradually increases within the stable region and like the longitudinal spacing depends on the Reynolds number. Experimental values of the mean relative spacing h/l vary from 0.19 to 0.3.

The vortex shedding results in oscillatory forces on the body both in the drag and lift direction. If there is a single vortex shedding frequency the force F_L in the lift direction can be approximated as

$$F_L(t) = |F_L| \cos(2\pi f_v t + \alpha) \tag{6.21}$$

where α is a phase angle. The lift force amplitude $|F_L|$ is normally expressed in terms of a lift coefficient C_L, which for a circular cylinder is defined as

$$C_L = \frac{|F_L|}{\frac{1}{2}\rho U_\infty^2 D} \tag{6.22}$$

In this case $|F_L|$ means the lift force amplitude per unit length.

There is large scatter in experimentally determined lift coefficients. The largest C_L-value reported by Sarpkaya & Isaacson (1981) was 1.35 for subcritical flow past a fixed circular cylinder. It should be noted that the vortex shedding is more or less uncorrelated along the cylinder axis. For subcritical flow past a fixed circular cylinder the correlation length is smaller than 5 times the diameter and for transcritical flow it may not be more than 1–2 times the diameter. The consequence of this is that the phase angle α in equation (6.21) varies along the cylinder axis. This means that the integrated lift force along the length of a cylinder will be small due to cancellation of force contributions from different cross-sections of the cylinder.

If there is a single vortex shedding frequency the force in the drag direction can be approximated as

$$F_D(t) = \bar{F}_D + A_D \cos(4\pi f_v t + \beta) \tag{6.23}$$

Here $\bar{F}_D$ is time independent and is the basis for the normally defined drag coefficients (see for instance Fig. 6.3). The amplitude A_D of the oscillatory part of the drag force is typically 20% of $\bar{F}_D$. We note from equations (6.21) and (6.23) that the oscillation frequency of the oscillatory part of the drag force is twice the oscillation frequency of the vortex shedding frequency and the lift force. The reason why the oscillation frequency is $2f_v$ is that a vortex is shed from the cylinder with a period $T_v/2$. The fact that this occurs from alternate sides does not matter for the drag force. However, for the lift force the force direction is influenced by which side of the cylinder the vortex is shed. Therefore the period of the lift force is T_v. Due to lack of correlation of vortex shedding along the cylinder axis, the phase angle β in equation (6.23) will vary strongly along the cylinder axis in the same way as α in equation (6.21).

The oscillatory forces due to vortex shedding may cause resonance oscillations of structures. This will be discussed later in the chapter.

CURRENT LOADS ON SHIPS

Empirical formulas are often used to calculate current forces and moments on a ship. The drag force F_1^c on the ship in the longitudinal direction will be mainly due to frictional forces. In the calculation of F_1^c one may generalize procedures followed in the estimation of ship resistance of a ship in still water. The Froude number $Fn = U_c/(Lg)^{\frac{1}{2}}(U_c = $ current velocity, $L = $ ship length) is so small that wave resistance can be totally neglected relative to viscous resistance. The following approximate formula may be used:

$$F_1^c = \frac{0.075}{(\log_{10} Rn - 2)^2} \tfrac{1}{2}\rho S U_c^2 \cos\beta \,|\cos\beta| \tag{6.24}$$

Here β is the angle between the current velocity and the longitudinal x-axis. (The positive x-direction is in the aft direction). S is the wetted surface of the ship and

$$Rn = \frac{U_c L \,|\cos\beta|}{v} \tag{6.25}$$

An alternative is to use a generalized Hughes (1954) formula. That means we replace $0.075/(\log_{10} Rn - 2)^2$ in equation (6.24) by $0.066(1 + k)$ $/(\log_{10} Rn - 2.03)^2$ where k is a form factor found from experiments. Typical values vary between 0.2 and 0.4 when $\beta = 0$. If the flow separates in the stern, k may be up to 0.8.

To evaluate the transverse viscous current force and current yaw moment on the ship one can use the cross-flow principle as long as the current direction is not close to the longitudinal axis of the ship. The cross-flow principle assumes that the flow separates due to cross-flow past the ship, that the longitudinal current components do not influence the transverse forces on a cross-section, and that the transverse force on a cross-section is mainly due to separated flow effects on the pressure distribution around the ship. This means we write the transverse current force F_2^c on the ship as

$$F_2^c = \tfrac{1}{2}\rho\left[\int_L dx C_D(x)D(x)\right]U_c^2 \sin\beta\, |\sin\beta| \qquad (6.26)$$

where the integration is over the length L of the ship. Further, $C_D(x)$ is the drag coefficient for cross-flow past an infinitely long cylinder with the cross-sectional area of the ship at the longitudinal coordinate x. $D(x)$ is the sectional draught.

The yaw moment F_6^c due to current flow is the sum of the Munk moment and the viscous yaw moment due to cross-flow. We can write

$$F_6^c = \tfrac{1}{2}\rho\left[\int_L dx\, C_D(x)D(x)x\right]U_c^2 \sin\beta\, |\sin\beta|$$

$$+ \tfrac{1}{2}U_c^2(A_{22} - A_{11})\sin 2\beta \qquad (6.27)$$

where A_{11}, A_{22} are added mass in surge and sway, respectively. The last term is the Munk moment and can be derived from non-separated potential theory. We will show the derivation of the Munk moment later by using a strip theory approach. The Munk moment is valid for any body shape.

We may note that the Munk moment and the viscous current moment have different angular dependence. This is illustrated in Fig. 6.10. The angular dependence of the transverse current force and the yaw moment predicted by equations (6.26) and (6.27) agrees reasonably well with experiment values in the vicinity of $\beta = 90°$. Fig. 6.11 presents theoretical and experimental values for transverse current force. 'Theoretical' is an ambitious word in this context. What we have done in the vicinity of 90° is to fit equation (6.26) to agree with the experimental value at $\beta = 90°$. However, we note that the angular dependence predicted by (6.26) is reasonable in the vicinity of 90°. For small angles of attack, i.e. β close to 0°, the transverse forces are estimated by considering the ship hull and the rudder to be lifting surfaces. (For more details about the physics see Newman, 1977.)

The results for the yaw moment are presented in Fig. 6.13. The moment is with respect to the z-axis defined in Fig. 3.2. A positive yaw

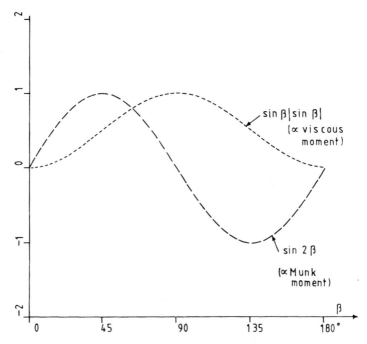

Fig. 6.10. Angular dependence of viscous yaw moment and Munk moment.
($\beta = 0°$ means current direction towards the stern of the ship.)

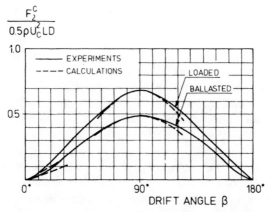

Fig. 6.11. Transverse current force F_2^c on the ship presented in Table 6.2 and
Fig. 6.12. (U_c = current velocity, L = ship length, D = draught,
β = current angle relative to the longitudinal direction of the ship.
$\beta = 0°$ means current direction towards the stern.) (Faltinsen et al.,
1979.)

Table 6.2. *Main particulars of a* 130 000 *DWT* (*dead weight tonne*) *ship*

| Parameter | Unit | Condition | |
		Ballast	Loaded
Length between perpendiculars	m	285.60	285.60
Beam	m	46.71	46.71
Depth	m	20.35	20.35
Draught fore	m	4.84	13.82
Draught aft	m	7.04	13.82
Draught mean	m	5.94	13.82
Displacement	tonnes	61 754.00	154 980.00
Centre of gravity, longitudinally from midships, positive aft	m	2.10	6.46
Centre of gravity, vertically from baseline	m	9.73	11.03
Metacentric height	m	21.50	8.97
Pitch/yaw radius of gyration	m	71.40	71.40
Roll radius of gyration	m	16.35	16.35
Natural pitch period	s	9.80	9.80
Natural roll period	s	9.40	12.80

moment means the current tends to rotate the ship to a smaller angle β. Again we have fitted the 'theory', i.e. equation (6.27), to agree with the experimental results at $\beta = 90°$. The angular dependence is now more complicated than for the transverse force, but we see a good agreement between 'theory' and experiments in the vicinity of 90°. The figure illustrates that the Munk moment and the viscous yaw moment are of equal importance. One way to see this is by noting the magnitude of the yaw moment at $\beta = 90°$. This is only due to viscous drag effects. This can be compared with the yaw moment at $\beta = 45°$, where the Munk moment has its maximum value and the viscous yaw moment is $1/\sqrt{2}$ times the value at $\beta = 90°$ according to the 'cross-flow' principle.

C_D-values for ship sections
In order to improve the predictions by equations (6.26) and (6.27) we need to know more about C_D-values. It is difficult to do this by theoretical means only. We will in the following text discuss what are the

important parameters influencing C_D. Important parameters are for instance free-surface effects, beam–draught ratio, bilge radius, bilge keel dimensions, Reynolds number and three-dimensional effects.

Free-surface effects

The free-surface tends to act as an infinitely long splitter plate. Hoerner (1965) refers to C_D-values for bodies with splitter plates of finite length in steady incident flow. The splitter plate causes a clear reduction of the drag coefficient.

A simple explanation of why the free-surface has an effect on the drag

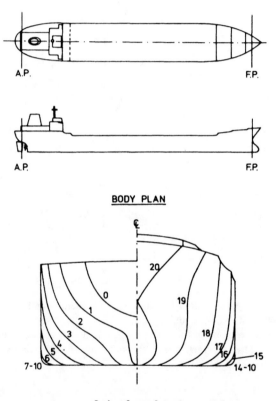

BODY PLAN

Body plan of tanker model

Fig. 6.12. Body plan of tanker model described in Table 6.2. F.P. = forward perpendicular. A.P. = aft perpendicular. The numbering of sections on the bottom part follows a standard numbering of ship cross-sections; it starts with section 1 on the aft part of the ship and ends with section 20 on the forward part of the ship. The figure shows only half of a cross-section. The other part is symmetric about the centre-line.

coefficient can partly be given by means of Fig. 6.14. The shed vorticity is represented by one single vortex of strength Γ, which is a function of time. To account for the free-surface effect one has to introduce an image vortex. This ensures zero normal velocity on the free-surface. If the splitter plate (free-surface) had not been there, instabilities would cause a Karman vortex street to develop behind the double body. The image vortex illustrated in Fig. 6.14 has a stronger effect on the motion of the

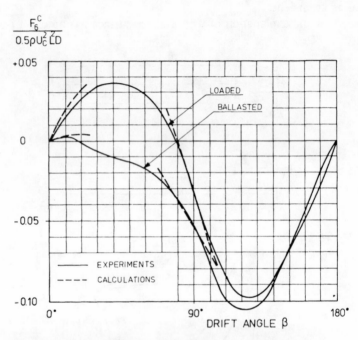

Fig. 6.13. Current yaw moment $F_6{}^c$ on the ship presented in Table 6.2 and Fig. 6.12. Notation defined in Fig. 6.11. Positive yaw moment implies rotation of the ship towards smaller β-angles. Moment is with respect to vertical axis through the centre of gravity of the ship. (Faltinsen *et al.*, 1979.)

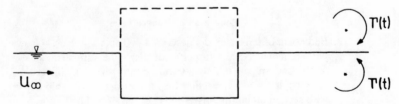

Fig. 6.14. Simple vortex system with an image flow above the free-surface so that the rigid free-surface condition is satisfied.

real vortex than the vortices in a Karman vortex street behind the double body have on each other. Since there is a connection between the velocities of the shed vortices and the force on the body we can understand why the free-surface influences the drag coefficient. In the case of oscillating flow at low amplitudes, which has relevance for eddy-making slow-drift damping, the eddies will stay symmetric for the double body without a splitter plate. This means the free-surface has little effect in this case.

Beam–draught ratio effects

Experimental results by Tanaka *et al.* (1982) show little effect of the height–length ratio on the drag coefficient for two-dimensional cross-sections of rectangular forms. One exception was for small height–length ratios. If one translates the results to midship cross-sections, it implies that the beam–draught B/D has small influence on the drag coefficient when $B/D > 0.8$.

Bilge radius effects

Experimental results by Tanaka *et al.* (1982) show a strong effect of the bilge radius r on the drag coefficient. Increasing the bilge radius means decreasing the drag coefficient. What we mean by strong effects can be seen by writing $C_D = C_1 e^{-kr/D} + C_2$, where C_1 and C_2 are constants of similar magnitude and D is the draught. As an example k may be 6.

Effect of laminar or turbulent flow

The classical results for a circular cylinder show that there is a critical Reynolds number. Below the critical Reynolds number the boundary layer is laminar. In the supercritical and transcritical range the boundary layer is turbulent. The consequence of this is quite different separation in the subcritical and transcritical Reynolds number ranges. A further consequence is a difference in drag coefficient. For a smooth cylinder the critical Reynolds number is at $2 \cdot 10^5$. By increasing the roughness of the cylinder surface, the critical Reynolds number will decrease. For marine structures one often has the situation that model tests have to be performed in the subcritical range, while the full-scale situation is in the transcritical range. However, when the separation occurs from sharp corners one would expect less severe scale effects.

Aarsnes (1984) (see also Aarsnes *et al.* 1985) has shown that the drag coefficient may be substantially different depending on laminar or turbulent separation. This is also evident from Delany & Sorensen's (1953) results. Aarsnes' results were for ship cross-sectional forms. The ship body plan is shown in Fig. 6.15. The results are presented in Fig. 6.16, both for subcritical and transcritical flow. Estimates for the drag coefficients for other cross-sections along the ship in subcritical flow are also shown in Fig. 6.16. In general there is a significant difference between the drag coefficient in subcritical and transcritical flow. The reason is that the flow separates more easily in subcritical flow. This is illustrated in Figs. 6.17 and 6.18 for the midship cross-section. In subcritical flow where the boundary layer flow is laminar, the flow separates at the 'leading' bilge. However, a turbulent boundary layer which occur in transcritical flow, can sustain a larger adverse pressure gradient without separating. This is the reason why there is no separation at the 'leading' bilge for transcritical flow. If separation occurs at both corners it seems as if the drag coefficient is roughly speaking twice the value compared to the case when separation occurs only at one corner.

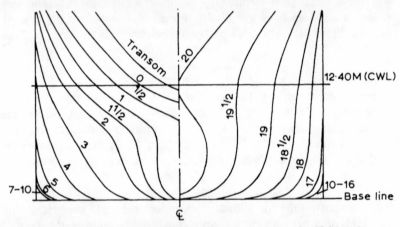

Fig. 6.15. Body plan of the ship examined in Fig. 6.16 (see also Table 6.3). Numbering of sections explained in Fig. 6.12.

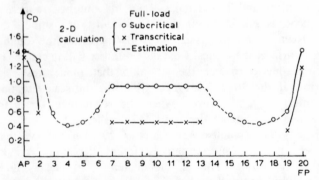

Fig. 6.16. Calculated and estimated drag coefficients C_D for two-dimensional cross-flow past cross-sections along the ship presented in Fig. 6.15. (Adapted from Aarsnes et al., 1985.)

Bilge keel effects

One important effect of bilge keels is that the separation points are fixed along the bilge keel and are not scale dependent. The length of bilge keels is typically half the ship length. A consequence of this is probably that ships with a bilge keel avoid the serious problems associated with scaling of transverse current force from model scale to full scale. Since the bilge keels are centered midships, we are more uncertain about the scaling of yaw current moment. The drag coefficient is not very sensitive to the breadth of the bilge keel.

Table 6.3. *Main particulars of ship presented in Fig.* 6.15

	Symbol	Unit	Model
Length between perpendiculars	L_{pp}	m	5.785
Beam	B	m	0.794
Draught	D	m	0.305
Block coefficient	C_{B}		0.84
Wetted surface area	S	m^2	7.096

Length scale ratio of ship to model: 40.62.

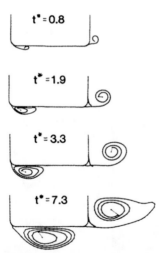

Fig. 6.17. Wake development for cross-section 10 of the ship presented in Fig. 6.15 (subcritical). The calculations are shown for different time instants t and are based on numerical calculations by Aarsnes (1984) with a thin free shear layer model. ($t^\star = U_c t/D$, U_c = current velocity, $t = 0$ corresponds to start time of the calculations, D = draught.) (Aarsnes, 1984.)

Three-dimensional effects

Aarsnes (1984) pointed out that three-dimensional effects at the ship ends will reduce the drag force compared to a pure strip theory approach. One way of taking this into account would be to use a reduced effective incident flow at ship ends as predicted in a qualitative way by Aarsnes. Physically, the reduced inflow velocity is due to the eddies at the ship ends (see Fig. 6.19). The effective reduced inflow can be translated into a reduction factor of the two-dimensional drag

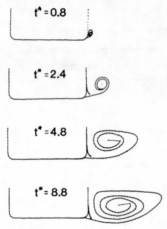

Fig. 6.18. Wake development for cross-section 10 of the ship presented in Fig. 6.15 (transcritical flow) (See also explanation to Fig. 6.17). (Aarsnes, 1984.)

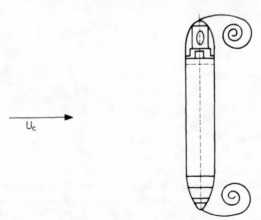

Fig. 6.19. Sketch of the vertical vortex system at ship ends present in cross-flow past the ship.

coefficients. The reason is that the forces on a cross-section are proportional to the square of the local inflow velocity, i.e. v^2, and that the drag coefficient is normalized by the square of the current velocity, i.e. U_c^2. The reduction factor is simply v^2/U_c^2. This reduction factor is presented in Fig. 6.20 and should be multiplied by the two-dimensional results in Fig. 6.16 to obtain the correct local two-dimensional solution. The local two-dimensional results can be added together by a strip-theory approach to find the three-dimensional results.

Munk moment

We will now derive the strip theory version of the Munk moment presented in equation (6.27). We write the velocity potential for non-separated flow as

$$\Phi = U_c x \cos \beta + U_c y \sin \beta + \phi \tag{6.28}$$

The perturbation potential ϕ due to the presence of the body satisfies the three-dimensional Laplace equation, the body boundary condition

$$\frac{\partial \phi}{\partial n} = -n_1 U_c \cos \beta - n_2 U_c \sin \beta \tag{6.29}$$

and the rigid free-surface condition. Here $\mathbf{n}$ is the normal to the ship surface with positive direction into the fluid. We can write

$$\mathbf{n} = (n_1, n_2, n_3) \tag{6.30}$$

If we use the strip theory concept we will neglect x-derivatives relative to y- and z-derivatives in the Laplace equation. This means ϕ satisfies a two-dimensional Laplace equation in the cross-sectional plane. The body boundary condition can be written as

$$\frac{\partial \phi}{\partial N} = -n_2 U_c \sin \beta \tag{6.31}$$

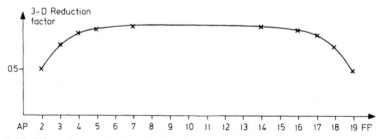

Fig. 6.20. Three-dimensional reduction factor of local drag coefficient due to the vertical vortex system at ship ends described in Fig. 6.19. (Adapted from Aarsnes *et al.*, 1985.)

where N is the two-dimensional normal in the cross-sectional plane. We can write

$$\phi = -U_c \sin \beta \, \phi_2 \qquad (6.32)$$

where ϕ_2 is the normalized two-dimensional sway velocity potential at zero-frequency. By 'zero'-frequency we mean a very small frequency. Due to the zero-frequency assumption ϕ_2 will satisfy the rigid free-surface condition. From ϕ_2 we can derive the added mass in sway. The 'zero-frequency' two-dimensional sway added mass for one cross-section can be written as

$$A_{22}^{(2D)}(x) = -\rho \int_{C(x)} \phi_2 n_2 \, ds \qquad (6.33)$$

where $C(x)$ is the wetted cross-sectional surface.

Let us now pursue the calculation of the current yaw moment. The pressure follows from Bernoulli's equation and can be written as (hydrostatic pressure is excluded)

$$p + \frac{\rho}{2} \left[\left(\frac{\partial \Phi}{\partial x} \right)^2 + \left(\frac{\partial \Phi}{\partial y} \right)^2 + \left(\frac{\partial \Phi}{\partial z} \right)^2 \right] = p_0 + \tfrac{1}{2} \rho U_c^2 \qquad (6.34)$$

Since the constant ambient pressure p_0 does not contribute to the force, we simply set it equal to zero in the following text. This means

$$p = -\rho U_c \cos \beta \frac{\partial \phi}{\partial x} - \rho U_c \sin \beta \frac{\partial \phi}{\partial y}$$

$$- \frac{\rho}{2} \left[\left(\frac{\partial \phi}{\partial x} \right)^2 + \left(\frac{\partial \phi}{\partial y} \right)^2 + \left(\frac{\partial \phi}{\partial z} \right)^2 \right]$$

$$= \frac{\rho}{2} U_c^2 \sin 2\beta \frac{\partial \phi_2}{\partial x} + \rho U_c^2 \sin^2 \beta \left\{ \frac{\partial \phi_2}{\partial y} - \tfrac{1}{2} |\nabla \phi_2|^2 \right\} \qquad (6.35)$$

The yaw moment can be written as

$$F_6 = \frac{\rho}{2} U_c^2 \sin 2\beta \int_L dx \, x \int_{C(x)} \frac{\partial \phi_2}{\partial x} n_2 \, ds$$

$$+ \rho U_c^2 \sin^2 \beta \int_L dx \, x \int_{C(x)} \left\{ \frac{\partial \phi_2}{\partial y} - \tfrac{1}{2} |\nabla \phi_2|^2 \right\} n_2 \, ds \qquad (6.36)$$

This expression can be further simplified. We will illustrate this by an example where the sum of the ship and the 'image' hull above the free-surface has circular cross-sections with radius $R(x)$ (see Fig. 6.21).

The velocity potential ϕ_2 can be written

$$\phi_2 = -\frac{R^2(x)}{r} \cos \theta \qquad (6.37)$$

This expression satisfies the body boundary condition $\partial\phi_2/\partial N = n_2$ which can be written $\partial\phi_2/\partial r = \cos\theta$. Further, it satisfies the rigid free-surface condition $\partial\phi_2/\partial z = 0$ on $z = 0$ and the two-dimensional Laplace equation for each cross-section of the ship. In evaluating the pressure on the body (see equation (6.35)) we need to know $\partial\phi_2/\partial x$, $\partial\phi_2/\partial y$ and $|\nabla\phi_2|^2$, as follows

$$\frac{\partial\phi_2}{\partial x} = -\frac{dR^2(x)}{dx}\frac{\cos\theta}{r}$$

$$\left.\frac{\partial\phi_2}{\partial y}\right|_{r=R(x)} = \cos 2\theta$$

$$|\nabla\phi_2|^2_{r=R(x)} = 4\left(\frac{dR}{dx}\right)^2\cos^2\theta + 1$$

When calculating the yaw moment we need to evaluate

$$\int_{C(x)}\left(\frac{\partial\phi_2}{\partial y} - \tfrac{1}{2}|\nabla\phi_2|^2\right)n_2\,ds$$

$$= \int_\pi^{2\pi}\left(\cos 2\theta - \frac{1}{2}\left(1 + 4\left(\frac{dR}{dx}\right)^2\cos^2\theta\right)\right)\cos\theta\, R\,d\theta$$

Since both $\partial\phi_2/\partial y$ and $\tfrac{1}{2}|\nabla\phi_2|^2$ are symmetric about the z-axis and n_2 is antisymmetric about the z-axis, it follows that the integral is zero. This is true for any cross-sectional surface that is symmetric about the z-axis. Hence the last term in equation (6.36) is zero.

For a ship with circular cross-section we can write the integral over

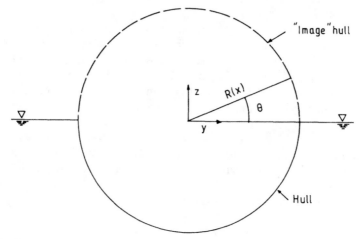

Fig. 6.21. Cross-section of a ship hull and image hull above the free-surface.

$C(x)$ in the first term of equation (6.36) as

$$\int_{C(x)} \frac{\partial \phi_2}{\partial x} n_2 \, ds = -\frac{dR^2(x)}{dx} \int_{\pi}^{2\pi} \cos^2 \theta \, d\theta = -\frac{dR^2(x)}{dx} \frac{\pi}{2}$$

By integrating by parts and using the fact that the cross-sectional area is zero at the ship ends it follows that

$$F_6 = 0.5 U_c^2 \sin 2\beta \int_L dx \, \rho \frac{\pi}{2} R^2(x) = 0.5 U_c^2 \sin 2\beta \, A_{22} \quad (6.38)$$

This result can be derived in a similar way for other cross-sectional shapes. According to slender body theory (strip theory) the added mass in surge A_{11} can be neglected relative to the added mass in sway A_{22}. Our result for the Munk moment is therefore consistent with the result in equation (6.27). We note that the result is independent of the origin of the coordinate system. The reason is that the derivation is based on potential theory and that there is no force acting on a body in infinite, inviscid and irrotational fluid (D'Alemberts paradox). In our case the fluid is not infinite in all directions. However, due to the rigid free-surface condition the problem of studying the flow around a ship in incident horizontal current is equivalent to studying the flow around a double-body in infinite fluid. The double-body consists of the ship hull and its mirror image above the free surface.

What we have shown in the derivation above is that the Munk moment part of the current yaw moment on a ship is due to the pressure part $p = -\rho U_c \cos \beta \, \partial \phi / \partial x$ where ϕ is the velocity potential $-U_c \sin \beta \, \phi_2$ due to the cross-flow past the ship.

We note from the derivation that the effect of flow separation is neglected. Strictly speaking this is incorrect. What we have done is to account for flow separation in only one part of the expression of the current yaw moment (see equation (6.27)). The reason why this viscous part is not sufficient to describe the current yaw moment is that the three-dimensional effects are not properly accounted for. The pressure part $-\rho U_c \cos \beta \, \partial \phi / \partial x$ that gives the Munk moment represents a three-dimensional effect because it involves the variation of the velocity potential ϕ in the longitudinal x-direction of the ship.

CURRENT LOADS ON OFFSHORE STRUCTURES

The Munk moment on a general body can be derived from a formula presented by Newman (1977). For a body moving with a constant translational velocity $\mathbf{U} = (U_1, U_2, U_3)$ in an infinite, inviscid and irrotational fluid we can express the moment $\mathbf{M}$ by

$$\mathbf{M} = -\rho \mathbf{U} \times \iint_{S_B} \phi \mathbf{n} \, ds \quad (6.39)$$

(see equation (99) on p. 136 of Newman, 1977). The surface normal direction $\mathbf{n} = (n_1, n_2, n_3)$ on the body surface S_B is assumed positive out of the fluid in Newman's derivation. The velocity potential ϕ in equation (6.39) can be written as

$$\phi = \sum_{i=1}^{3} U_i \phi_i \tag{6.40}$$

where ϕ_i satisfies the body boundary condition

$$\frac{\partial \phi_i}{\partial n} = n_i \qquad i = 1, 2, 3 \text{ on } S_B \tag{6.41}$$

By using the fact that the added mass coefficient A_{ji} can be written

$$A_{ji} = \rho \iint_{S_B} \phi_i n_j \, ds \tag{6.42}$$

we can now write equation (6.39) as

$$\mathbf{M} = -\mathbf{U} \times \left[\sum_{i=1}^{3} U_i \left[\sum_{j=1}^{3} A_{ji} \mathbf{e}_j \right] \right] \tag{6.43}$$

where $\mathbf{e}_j$, $j = 1, 2, 3$ are unit vectors along the x-, y- and z-axis, respectively. By multiplying out the cross-product we find

$$\mathbf{M} = -\mathbf{e}_1 \left(U_2 \sum_{i=1}^{3} U_i A_{3i} - U_3 \sum_{i=1}^{3} U_i A_{2i} \right)$$

$$- \mathbf{e}_2 \left(U_3 \sum_{i=1}^{3} U_i A_{1i} - U_1 \sum_{i=1}^{3} U_i A_{3i} \right)$$

$$- \mathbf{e}_3 \left(U_1 \sum_{i=1}^{3} U_i A_{2i} - U_2 \sum_{i=1}^{3} U_i A_{1i} \right) \tag{6.44}$$

For instance, we can apply this formula to current yaw moment on a ship. Observed from the ship the translational velocity, $\mathbf{U}$, is the same as an incident current velocity. Then we can write $U_1 = U_c \cos \beta$, $U_2 = U_c \sin \beta$, $U_3 = 0$. From equation (6.44) it follows that the yaw moment is

$$-\mathbf{e}_3 U_c^2 \cos \beta \sin \beta (A_{22} - A_{11}) \tag{6.45}$$

This is the same as the Munk moment part in equation (6.27) if we recognize that the x-, y- and z-directions defined here are opposite the x-, y- and z-directions used in equation (6.44).

We can apply equation (6.44) to any offshore structure in a current. However, we should realize that we have assumed infinite, inviscid and irrotational fluid. As pointed out in the last section this means that there is no hydrodynamic force on the body. This is the reason why equation (6.44) does not depend on the position of the origin of the coordinate system. The added mass coefficients in equation (6.44) can for instance be calculated by the source technique presented in chapter 4.

Another application of equation (6.44) could be to study the behaviour of a submarine in steady forward motion in infinite fluid. This means that $U_2 = U_3 = 0$ and we find

$$\mathbf{M} = \mathbf{e}_2 U_1^2 A_{31} - \mathbf{e}_3 U_1^2 A_{21} \tag{6.46}$$

Since a submarine has port–starboard symmetry it follows that $A_{21} = 0$ and since A_{31} is generally non-zero it means that the submarine experiences a pitching moment. The sign of the moment depends on the sign of the coupled added mass coefficient A_{31} between heave and surge. This effect is also important for SWATH (small water plane area twin hulls) ships.

We can generalize the way viscous current loads were calculated for a ship (see equations (6.26) and (6.27)) to offshore structures that consist of slender structural parts. Examples of the latter are columns and pontoons of semi-submersibles and of tension leg platforms (TLP). Other examples would be risers or cables. This means we decompose the current velocity into one component in the longitudinal direction and one component in the cross-flow direction of a slender structural part. We assume that the longitudinal current velocity component causes only shear forces as shown in equation (6.24) for longitudinal drag forces on a ship. The cross-flow component causes flow separation. The force on a cross-section (strip) due to flow separation can be calculated in a similar way to equation (6.26). Thus, the mean force per unit length can be written as (see Fig. 6.22)

$$F_{\mathrm{D}} = \frac{\rho}{2} C_{\mathrm{D}} D U_{\mathrm{N}}^2 \tag{6.47}$$

$$\bar{F}_{\mathrm{L}} = \frac{\rho}{2} \bar{C}_{\mathrm{L}} D U_{\mathrm{N}}^2 \tag{6.48}$$

Here D is a characteristic cross-sectional length such as the diameter for a circular cylinder. Further, F_{D} is the mean force in the same direction as the cross-flow component U_{N} of the current velocity and $\bar{F}_{\mathrm{L}}$ is the mean force orthogonal to F_{D} in the cross-flow plane. $\bar{F}_{\mathrm{L}}$ is zero for a single body in infinite fluid when the body is symmetric about an axis parallel to the direction of U_{N}. Based on equations (6.47) and (6.48) we can calculate total current force and moment on the structure. This is similar to that shown in equations (6.26) and (6.27) for a ship. The drag and lift coefficients C_{D} and $\bar{C}_{\mathrm{L}}$ have to be empirically determined. Important parameters influencing C_{D} and $\bar{C}_{\mathrm{L}}$ are discussed in the section on current loads on ships. In addition it should be noted that there may be hydrodynamic interaction between structural parts. For instance, if a cylinder is placed in the wake behind another cylinder it will experience a smaller incident velocity and therefore a smaller drag coefficient if the free stream is used to normalize the drag coefficient. This can be illustrated by using a theoretical expression for the wake velocities far behind a two-dimensional body. According to Schlichting (1979: pp. 742–3), we can write the mean wake velocity component u in the direction of the incident velocity as

$$\frac{u}{U_\infty} = 1 - 0.95\left(\frac{x}{C_D D}\right)^{-\frac{1}{2}} e^{-\frac{1}{4}\eta^2} \tag{6.49}$$

where

$$\eta = y(0.0222 C_D D x)^{-\frac{1}{2}} \tag{6.50}$$

Here U_∞ is the incident velocity.

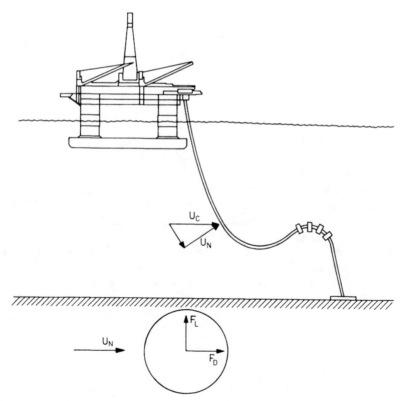

Fig. 6.22. Definition of force components on a cross-section of a slender structural member. (U_c = incident current velocity.)

The x-axis is assumed parallel to the incident flow velocity direction and the y-axis is orthogonal to the x-axis. The origin of the coordinate system is at the geometrical centre of the body. Further, C_D is the drag coefficient and D is the frontal area of the body per unit length projected on the y-axis or another characteristic length used in describing the drag forces (see equation (6.47)). The solution described by equation (6.49) agrees very well with experimental data when $x/(C_D D) > 50$. Below this value it is not clear if equation (6.49) can be used for practical purposes. However, it should be remembered that the velocity difference $U_\infty - u$ is assumed to be small compared with the free stream velocity U_∞. If we consider two circular cylinders in tandem with a large distance l between the centres of the cylinders (see Fig. 6.23) we can use equation (6.49) to calculate the drag coefficient $C_D^{(2)}$ on the cylinder in the wake of the other cylinder. The square of the incident velocity acting on the wake cylinder is then approximately

$$u^2 \approx U_\infty^2\left(1 - 2 \cdot 0.95\left[\frac{C_D D}{l}\right]^{\frac{1}{2}}\right) \tag{6.51}$$

Equation (6.51) gives the following result

$$C_D^{(2)} = \frac{\frac{\rho}{2}C_D D u^2}{\frac{\rho}{2}U_\infty^2 D} = C_D - 1.9 C_D^{\frac{3}{2}}\left(\frac{D}{l}\right)^{\frac{1}{2}} \tag{6.52}$$

If $C_D = 1.0$ we find from equation (6.52) that $C_D^{(2)}$ is respectively 0.58, 0.7 and 0.81 for $l/D = 20$, 40 and 100. Strictly speaking we should not apply the formula when $l/D = 20$ and 40.

The above procedure can be generalized to several bodies, where the wake from one body influence other bodies. Zdravkovich (1985) has given a survey of results for interaction of pipe clusters in steady incident flow. We will refer to some of his results for two cylinders. He defines three basic interference categories. That is *proximity interference* (P), *wake interference* (W) and *no interference*. The first kind means that the examined cylinder is not in the wake of the other cylinder. The different

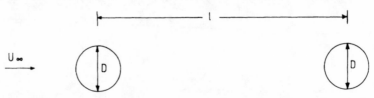

Fig. 6.23. Definition of parameters for two cylinders arranged in tandem position.

regions are shown in Fig. 6.24(a). It is assumed that the two cylinders have the same radius. Only one reference cylinder is shown. This cylinder is for instance downstream of the other cylinder if $x/D < 0$ in Fig. 6.24(a).

Examples of the flow picture that occurs in the different regions are shown in Fig. 6.24(b). For example, in a side by side arrangement we see only one *single* vortex street when $1 < y/D < 1.1$ to 1.2. When $2.7 < y/D < 4 - 5$ we see that the two vortex streets mirror each other. One way to estimate that there is little influence between two cylinders in a side by side arrangement when $y/D > \approx 5$ is to use the analytical

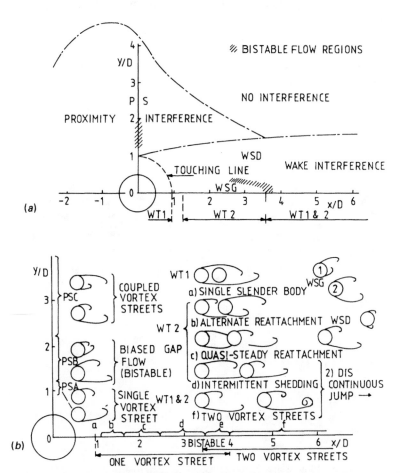

Fig. 6.24 (a). Regions of flow interference categories, (b) interference flow regimes (schematic). PSA, PSB and PSC are different parts of the area described as PS in (a). WT1, WT2, WSD and WSG are explained in (a). (Zdravkovich, 1985.)

potential flow solution around a circular cylinder at the position of the other cylinder.

In a tandem arrangement we see that the vortex street behind the cylinders is formed from the free-shear layer detached from the first cylinder when $1 \leqslant x/D < 1.2-1.8$. When $1.2-1.8 \leqslant x/D < 3.4-3.8$ the free-shear layer that separates from the first cylinder, re-attaches to the second cylinder. Fig. 6.25 shows a photo of vortex formation around two vertical cylinders in a tandem arrangement in steady incident flow. The cylinders were towed in a model basin. The visualization was done by placing confetti on the free-surface and taking photographs. The photo illustrates that a vortex from the upstream cylinder can be trapped between the two cylinders.

Zdravkovich (1985) presented results for mean lift, drag and Strouhal number both for the subcritical and supercritical flow regimes. If the cylinders are close he shows that the drag may be negative on the second cylinder in a tandem position. This may occur if $x/D < \approx 4$. The results are dependent on the Reynolds number, the roughness ratio and the number of cylinders. The minimum drag coefficient that Zdravkovich reports is ≈ -0.6.

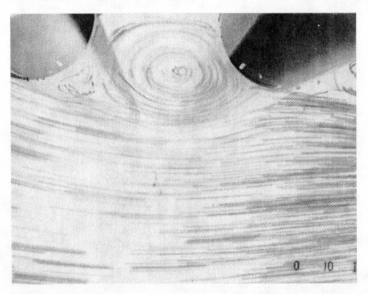

Fig. 6.25. Flow visualization of vortex formation around two cylinders in tandem arrangement in steady incident flow. The photo illustrates that a vortex from the upstream cylinder can be trapped between the two cylinders.

WIND LOADS

Wind loads can be estimated in a similar way to current loads. Empirical or experimental data is necessary. For instance, Isherwood (1973) has presented drag coefficients for passenger ships, ferries, cargo ships, tankers, ore carriers, stern trawlers and tugs, and Aquirre & Boyce (1974) have estimated wind forces on offshore drilling platforms.

VORTEX-INDUCED RESONANCE OSCILLATIONS

The oscillatory forces due to vortex shedding may cause resonance problems. This is well known in many fields of engineering. We will illustrate vortex-induced oscillations by two examples.

One case is a moored loading buoy in a uniform current of velocity U_c. The buoy is a vertical cylinder of length L and diameter D, where $L/D \gg 1$. The vortex-shedding period T_v can be found from the Strouhal number St, i.e.

$$T_v = \frac{1}{\text{St}} \cdot \frac{D}{U_c} \qquad (6.53)$$

Let us assume the current direction to be parallel to the x-axis, $D = 20$ m and $U_c = 1$ m s^{-1}. The corresponding Reynolds number is $\approx 2 \cdot 10^7$. This means St ≈ 0.25 (see Fig. 6.26) and $T_v = 80$ s. If T_v is in the vicinity of the natural period for sway oscillation, the consequence is large vortex-induced sway oscillations and drag-forces. Both the resulting horizontal

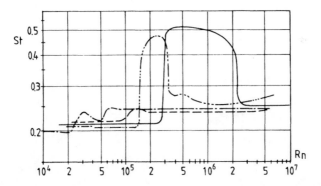

Fig. 6.26. Strouhal number St of rough circular cylinders in steady incident flow for different surface roughness values k/D (k = average height of surface roughness, D = cylinder diameter, St $= f_v D/U_\infty$, f_v = vortex shedding frequency, U_∞ incident flow velocity, Rn = $U_\infty D/\nu$.) ——, smooth; —$\cdots$—, $k/D = 7.5 \cdot 10^{-4}$; — —, $k/D = 3 \cdot 10^{-3}$; —$\cdot$—, $k/D = 9 \cdot 10^{-3}$; ----, $k/D = 3 \cdot 10^{-2}$. (Achenbach & Heinecke, 1981.)

excursions of the anchor-lines and the drag-forces may be of importance in the design of the mooring system.

Another example is a riser in a current. The vortex-induced oscillations may cause elastic resonance oscillations. We will illustrate this with an example and consider a family of risers for different water-depths up to 2000 m (see Fig. 6.27). The top tension was kept constant equal to 1250 kN for all risers. For increased water depths the end geometries were unchanged while the buoyancy parts were increased with length. The submerged weight of this part of the riser was close to zero. The natural periods for different eigenmodes are shown as a function of water depth in Fig. 6.28. Let us as an example consider a current velocity of 0.8 m s^{-1}. For the buoyancy part $D = 1$ m. This means a Reynolds number of $\approx 8 \cdot 10^5$. For a smooth cylinder this implies the supercritical flow regime, where there is a vortex-shedding frequency spectrum. However, in reality there is likely to be marine growth. The higher the roughness ratio is, the lower the critical Reynolds number is (see Fig. 6.3). With a roughness height of 7 mm, the critical Reynolds number is at $\approx 6 \cdot 10^4$, and $Rn \approx 8 \cdot 10^5$ will mean transcritical flow. Let us for simplicity set $St = 0.25$. According to equation (6.53) $T_v = 5$ s. Depending on the water depth (see Fig. 6.28), this could correspond to an eigenperiod and cause resonance oscillations. On the other hand if $T_v = 5$ s is not in the vicinity of an eigenperiod for one particular water depth, a different realistic current velocity is likely to cause resonance oscillations.

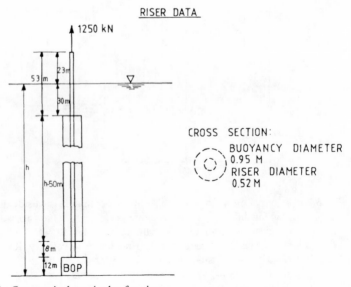

Fig. 6.27. Geometrical particular for riser.

The hydroelastic oscillations have several peculiarities which do not happen for a linear system. We will try to show this by referring to experimental results by Feng (1968) with a lightly damped circular cylinder in a current. Results for transverse oscillation amplitude A, vortex shedding frequency f_v and cylinder frequency f_c are presented in Fig. 6.29 as a function of the reduced velocity $U_R = U_c/(f_n D)$. Here f_n is the natural frequency of oscillation of the cylinder. When the vortex shedding frequency f_{vo} for a stationary cylinder approaches the natural frequency f_n from below ($U_R \approx 5$ in Fig. 6.29), the vortex shedding frequency becomes nearly identical to the natural frequency of the cylinder. Furthermore, the frequency of oscillation of the cylinder, f_c, nearly coincides with its natural frequency. As U_R increases towards 6.2, the amplitude of the transverse oscillation A increases rapidly while f_v/f_n and f_c/f_n remain nearly constant at a value slightly less than unity. As U_R increases further, the amplitude drops abruptly as illustrated in Fig. 6.29. For $U_R \geqslant \approx 7$ the frequency of vortex shedding, f_v, jumps to its 'stationary-cylinder' value, i.e., $f_v = f_{vo}$. However, the cylinder continues to oscillate at the natural frequency with very small amplitudes. The region $5 \leqslant U_R \leqslant 7$ where lock-in of the vortex shedding frequency to the natural frequency of the cylinder occurs, is variously referred to as the synchronization region, lock-in region, capture region, or simply as

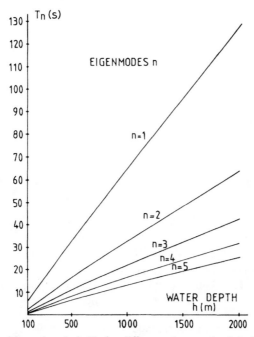

Fig. 6.28. Natural periods T_n for different eigenmodes 'n' of riser in Fig. 6.27.

the resonant region. There does not exist any engineering tool based on sound physical principles that can predict what is happening in the lock-in region. However, Sarpkaya & Shoaff (1979) have presented interesting results based on the discrete vortex method.

Let us mention some consequences of lock-in

     A. Correlation length increases
     B. Vortex strength increases
     C. Vortex shedding frequency locks onto the natural frequency
     D. Increasing amplitude means increasing the band width of lock-in
     E. Oscillations are self-limiting (maximum relative amplitude $A/D \approx 1$)
     F. In-line drag force increases

One could argue that there is a connection between the increase in correlation length (point A) and the vortex shedding frequency locking onto the natural frequency (point C) by saying that the vortex shedding

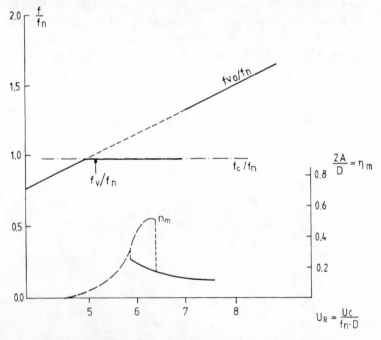

Fig. 6.29. Experimental results for self-excited transverse oscillations of a cylinder. (f_n = natural frequency of oscillation of the cylinder, f_c = frequency of oscillation of the cylinder, f_{vo} = vortex shedding frequency for fixed cylinder, f_v = vortex shedding frequency, U_c = current velocity, A = amplitude of cylinder oscillations ---- f_{vo}/f_n, -·- f_c/f_n, —— f_v/f_n). (Adapted from Feng, 1968.)

must be correlated in phase with the cylinder oscillations and these cannot vary rapidly along the length due to structural constraints.

There is also a connection between the increase in vortex strength (point B) and the increase in drag forces (point F). Using the Blasius theorem (Sarpkaya & Shoaff, 1979), we can show how the drag and lift force are related to the vorticity and its velocity.

Blevins (1977), Griffin et al. (1975) and Sarpkaya (1978) have analysed the maximum amplitude A_{max} of transverse oscillation in the lock-in range for both elastically-mounted and flexible cylinders. Sarpkaya wrote the maximum amplitude for any mode shape as

$$(A/D)_{max} = 0.32\gamma/[0.06 + \Delta_r^2]^{\frac{1}{2}} \tag{6.54}$$

in which γ is given as

$$\gamma = \psi_{max}(y/L)\left[\frac{\int_0^L \psi^2(y)\,dy}{\int_0^L \psi^4(y)\,dy}\right]^{\frac{1}{2}} \tag{6.55}$$

Here L is the length of the cylinder and $\psi(y)$ is the mode shape. The reduced damping Δ_r is defined by

$$\Delta_r = (2\pi\,St)^2\frac{2m\zeta_s}{\rho D^2} \tag{6.56}$$

where ζ_s is the fraction of the structural damping to critical damping, St is the Strouhal number for the non-oscillating cylinder, ρ is the fluid density and m is the sum of structural mass and added mass per unit length.

For a spring-supported cylinder it follows that $\gamma = 1.0$. By using equation (6.54) and setting $\Delta_r = 0$ we find that $(A/D)_{max} = 1.31$. If we want to apply the above formulas to risers, we need to calculate the natural frequencies and eigenmodes by a standard computer program or use the simplified solutions of Sparks (1980). For marine applications Δ_r will have a small influence in equation (6.54). If we totally neglect it and approximate the mode shape by a sinusoidal form we find that $\gamma = 1.16$ and $(A/D)_{max} = 1.51$. Having determined the amplitude of oscillations of the riser and the mode shape we can easily calculate stresses due to the vortex-induced oscillations.

A very important consequence of the vortex-induced oscillations is that the drag force increases. Skop et al. (1977) proposed an empirical formula for a rigid circular cylinder. They write for $W_R > 1$

$$C_D/C_{D0} = 1 + 1.16(W_R - 1)^{0.65} \tag{6.57}$$

where

$$W_R = (1 + 2A/D)f_n/f_{v0} \tag{6.58}$$

A cruder approximation is $C_D/C_{D0} = 1 + 2(A/D)_{max}$. This formula can be interpreted as saying that there is an apparent projected area $D + 2A$ due to the oscillating cylinder.

We should be aware of the effect of roughness and Reynolds number on vortex-induced oscillations. Transverse oscillations decay in the critical region both for smooth and rough cylinders.

Griffin (1985) has discussed the effect of current shear on vortex shedding. He says that moderate shear levels of practical importance ($\bar{\beta} \approx 0.01$ to 0.03) do not seem to decrease the likelihood of vortex-induced vibrations for flexible cylinders. Experience with long cables in the ocean indicates that the vibration level is decreased. The 'steepness' parameter $\bar{\beta}$ is defined as

$$\bar{\beta} = \frac{D}{V_{ref}} \frac{d\bar{V}}{dz} \tag{6.59}$$

Here $d\bar{V}/dz$ is the vertical velocity gradient, D is the cylinder diameter for a circular cylinder and $\bar{V}_{ref}$ is a reference velocity, e.g. the mid-span incident velocity.

All the cases we have discussed are due to steady incident flow. However, vortex-induced oscillation may also occur in ambient harmonic oscillatory flow. We should also note the possibility of in-line oscillations. The reason is the oscillatory behaviour of the drag-force. In subcritical flow the drag-force on a fixed, rigid cylinder consists of the sum of mean drag-force and a harmonic oscillatory force of frequency twice the vortex shedding frequency. The in-line oscillations are significantly less than the transverse oscillations.

For a marine riser in waves and current we should note that the lock-in cannot take place on the whole riser at the same time. This is due to the attenuation of the waves with depth.

It is possible to suppress vortex-induced oscillations by various types of spoilers or by increasing the damping. We may also try to avoid the critical reduced velocities which cause vortex-induced oscillations. This means the reduced velocity $U_R = U_c/(f_n D)$ should not be in a range from five to seven.

A common device for suppression of vortex-induced oscillations is a helical strake (see Fig. 6.30). For a helical strake to be optimal it is said that one should use three spirals, set the pitch equal to five times the diameter (D) and choose a fin height of 0.1–$0.12D$.

Negative effects of the helical strakes are that marine growth reduces the effect and the fins cause an increased drag ($C_D = 1.3$ at $Rn = 5 \cdot 10^6$). There may also be handling problems associated with them.

GALLOPING

Galloping is a different phenomenon from 'lock-in'. 'Lock-in' happens when the vortex shedding frequency is close to a natural frequency for the structure and can in a simplistic way be understood as resonance oscillations. Galloping is instability oscillations of the structure that are significant for higher reduced velocities U_R than those when lock-in occurs. Let us explain how galloping can occur by considering an example with a long cylinder of constant cross-section. We assume two-dimensional flow in the cross-sectional plane. We define a coordinate system (x, y) as in Fig. 6.31. The incident flow is steady with velocity U_c in the positive x-direction. The body is symmetric about the y-axis and assumed to have a resonance frequency f_n when it oscillates in the y-direction. A fundamental assumption is that the fluid forces are quasi-steady. Oscillating vortex shedding forces do not matter. This is approximately correct if $U_R > 10$ (see Fig. 6.29). Galloping motion will occur if the hydrodynamic forces cause a sufficiently large negative damping of the transverse oscillations. We will show this in more detail.

The equation of motion for the transverse oscillation η_2 of the cylinder

Fig. 6.30. Helical strake used as a device for suppression of vortex-induced oscillations.

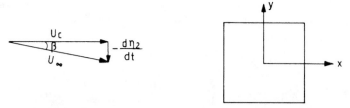

Fig. 6.31. Cross-sectional plane of a long cylinder used in discussion of galloping motions. (U_c = incident current velocity, U_∞ = incident velocity seen from a reference frame following the motion η_2 of the cylinder.)

can formally be written

$$(M + A_{22})\frac{\mathrm{d}^2\eta_2}{\mathrm{d}t^2} + 2\zeta_s(M + A_{22})\omega_n\frac{\mathrm{d}\eta_2}{\mathrm{d}t} + C_{22}\eta_2 = F_y \qquad (6.60)$$

where $\omega_n = 2\pi f_n = [C_{22}/(M + A_{22})]^{\frac{1}{2}}$ is the natural circular frequency of oscillation of the structure. The restoring force $-C_{22}\eta_2$ may be due to linear springs attached to the body. Further, ζ_s is the non-hydrodynamic damping as a fraction of the critical damping. The transverse force F_y in equation (6.60) is found by a quasi-steady approach. From a reference frame fixed to the body one sees an incident flow with direction

$$\beta = -\mathrm{arctg}\left[\frac{\mathrm{d}\eta_2}{\mathrm{d}t}\Big/U_c\right] \approx -\frac{\mathrm{d}\eta_2}{\mathrm{d}t}\Big/U_c \qquad (6.61)$$

relative to the x-axis. Based on experimental data we assume that a drag coefficient

$$C_y = \frac{F_y}{\dfrac{\rho}{2}U_\infty^2 A} \qquad (6.62)$$

is given as a function of β. This is similar to the data presented in Fig. 6.11 for transverse current forces on a ship. U_∞ means the ambient fluid velocity and A is the projected area of the body on the x-axis. By a Taylor expansion about $\beta = 0$ we find that

$$F_y \approx \beta\frac{\partial C_y}{\partial\beta}\Big|_{\beta=0}\frac{\rho}{2}U_c^2 A$$

By substituting this into equation (6.60) we find that

$$(M + A_{22})\frac{\mathrm{d}^2\eta_2}{\mathrm{d}t^2} + \left(2\zeta_s(M + A_{22})\omega_n + \frac{\rho}{2}U_c A\frac{\partial C_y}{\partial\beta}\Big|_{\beta=0}\right)\frac{\mathrm{d}\eta_2}{\mathrm{d}t}$$
$$+ C_{22}\eta_2 = 0 \quad (6.63)$$

When the damping coefficient of this equation is negative, the η_2 motion will be unstable. This means that a small initial disturbance of the body will increase exponentially with time. Unstable motions are what we call galloping. According to equation (6.63) this will occur if

$$\frac{\partial C_y}{\partial\beta}\Big|_{\beta=0} < -\frac{8\pi}{U_R}\frac{(M + A_{22})}{\rho AD}\zeta_s \qquad (6.64)$$

For a circular cylinder $\partial C_y/\partial\beta|_{\beta=0} = 0$. This means that galloping does not occur. In order to predict the amplitudes of galloping we have to simulate equation (6.60) with F_y as a function of the instantaneous value of the transverse velocity $\mathrm{d}\eta_2/\mathrm{d}t$. Parkinson (1985) gave a discussion as to how to combine galloping predictions with vortex-induced oscillations.

Overvik (1982) (see Fig. 6.32) saw galloping clearly in some of his tests for riser bundles. It happened typically for $U_R > 10$. Galloping motions may also occur for ships at a single point mooring system or for a ship which is towed (see exercise 6.3).

The galloping motions are not self-limited as are 'lock-in' oscillations. Actually at high velocities the amplitude of oscillation is proportional to the incident velocity. This also seems to be the case in Overvik's tests (see Fig. 6.32). On the other hand there are limitations on how large U_R can be in Fig. 6.32 in practical cases. We will discuss this. Let us use Fig. 6.28 which shows the natural periods for the structure shown in Fig. 6.27. Galloping will not occur for this riser, but let us use the natural periods as indicative of the natural periods for a riser bundle where problem may occur. We can write $U_R = T_n U_c/D$ and assume $U_c = 1 \, \text{m s}^{-1}$ and $D = 1 \, \text{m}$. From Fig. 6.28 we note that the highest natural period T_n is less than 10 s when the water depth is 100 m. This means that $U_R < 10$. Experience shows that there will not be any large amplitude due to galloping motion when $U_R < 10$. However, for 500 m water depth the U_R-value corresponding to the first eigenmode is larger than 30. This means the possibility of galloping should be considered.

EXERCISES

6.1 Current forces and flow through fishing nets

Fig. 6.33 shows a quadratic part of a cage. The netting panel has a mesh size $\lambda = 0.02 \, \text{m}$ and a twine diameter $D_i = 0.002 \, \text{m}$.

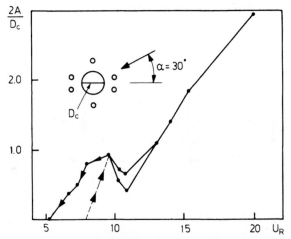

Fig. 6.32. Transverse double amplitude $2 \cdot A$ versus reduced velocity U_R for case C3, of Overvik's (1982) model tests. ($U_R = U_c/(f_n D_c)$, $U_c =$ current velocity, $f_n =$ natural frequency of oscillations of the cylinder, D_c is defined in the figure.)

(a) Start with equation (6.49), that describes the velocity distribution in the wake behind a cylinder. Show that the x-component u of the velocity in the point (x, y, z) behind the netting panel in Fig. 6.33 can be approximated as

$$\frac{u}{U_{\infty}} = 1.0 - 0.95 \left(\frac{x}{C_{\mathrm{D}}D} \right)^{-\frac{1}{2}} \left(\sum_{i=1}^{N_{i}} \exp\left(\frac{-(y-y_{i})^{2}}{0.0888C_{\mathrm{D}}Dx} \right) \right.$$

$$\left. + \sum_{j=1}^{N_{i}} \exp\left(-\frac{(z-z_{j})^{2}}{0.0888C_{\mathrm{D}}Dx} \right) \right) \qquad (6.65)$$

(b) Confirm that the velocity is approximately constant in transverse direction behind the netting panel.
Discuss how the current velocity will decay if there are several netting panels placed after each other as shown in Fig. 6.34.

(c) The current force on a netting panel can be calculated by the formula

$$F_{\mathrm{D}} = 0.5\rho C_{\mathrm{D}}U_{\mathrm{L}}^{2}A$$

($C_{\mathrm{D}} = 0.204$, $A = 1 \text{ m}^2$, upstream velocity $U_{\infty} = 1 \text{ m s}^{-1}$, $U_{\mathrm{L}} = $ local incident current-velocity ahead of a netting panel.) Justify that the drag coefficient is the right order of magnitude. Find the total current force on four equal netting

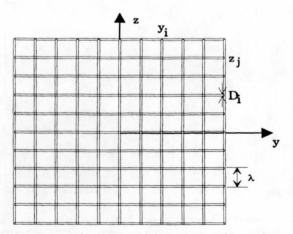

Fig. 6.33. Quadratic part of a cage used in exercise 6.1 ($\lambda = $ mesh size, $D_i = $ twine diameter).

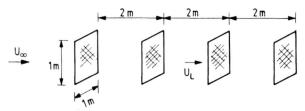

Fig. 6.34. Definition of netting panels used in exercise 6.1.

panels placed in the wake of each other. The distance between each netting panel is 2 m (see Fig. 6.34). (Answer: 313.4 N)

6.2   Vortex induced oscillations of a riser in a current

Consider a vertical rigid riser. The differential equation that describes the linear behaviour of a riser can be written as (Sparks, 1980)

$$EI \frac{d^4y}{dx^4} - T(x)\frac{d^2y}{dx^2} - W\frac{dy}{dx} + (m + A_{22})\frac{d^2y}{dt^2} = f(x) \qquad (6.66)$$

E – Young's modulus of elasticity

I – Inertia (flexural)

y – Horizontal offset from vertical through riser foot

x – Height above riser foot

T – Effective tension (calculated from consideration of the apparent weight of riser plus contents)

W – Linear apparent weight per unit length (riser plus contents). In order to get the true weight of the contents one has to add the weight of the content and $\rho g V$ where V is the displaced volume of the segment

m – Mass per unit length

A_{22} – Added mass in direction y per unit length

$f(x)$ – Lateral external load per unit length normally calculated by Morison's equation (see equation (3.34)).

Sparks show that rigidity has a negligible effect on riser curvature except for limited zones close to the ends. We will therefore neglect it in the following analysis.

Our purpose is to set up expressions for the eigenmodes and natural frequencies. This means we shall set $f(x) = 0$. The equation we shall study is therefore

$$T(x)\frac{d^2y}{dx^2} + W\frac{dy}{dx} - (m + A_{22})\frac{d^2y}{dt^2} = 0 \qquad (6.67)$$

We will consider a near vertical riser without buoyancy element. We can then write

$$T(x) = T_B + Wx \tag{6.68}$$

where T_B is effective tension at the riser foot. Sparks points out that the second term in equation (6.67) is negligible. It would therefore not make any significant difference if we write (6.67) as

$$(T_B + Wx)\frac{d^2y}{dx^2} + \frac{1}{2} W \frac{dy}{dx} - (m + A_{22})\frac{d^2y}{dt^2} = 0 \tag{6.69}$$

The eigenmodes and natural frequencies that satisfy equation (6.69) and boundary conditions $y = 0$ at $x = 0$ and $x = L$ (L = length of riser) are

$$\psi_n(x)e^{-i\omega_n t} = \sin(z_x - z_B)e^{-i\omega_n t} \qquad n = 1, 2, \ldots \tag{6.70}$$

where

$$z_x = \frac{2\omega_n}{W}[(m + A_{22})(T_B + Wx)]^{\frac{1}{2}} \tag{6.71}$$

$$T_n = \frac{2\pi}{\omega_n} = \frac{4L}{n}\frac{(m + A_{22})^{\frac{1}{2}}}{\sqrt{T_t} + \sqrt{T_B}} \tag{6.72}$$

Here T_t is the top tension. We should note that $WL = (T_t - T_B)$.

(a) Consider the riser presented in Fig. 6.27. Try to use equation (6.72) as an approximate solution for the natural periods and compare the results with Fig. 6.28.

(b) Assuming uniform current along a riser, use equations (6.70)–(6.72) to describe the eigenmodes and natural periods. Set up a formula for the maximum amplitude A_{max} of transverse oscillations due to vortex-induced oscillations (see equation (6.54)). The expressions should be in terms of main characteristics of the riser and the current velocity. Choose dimensions and apply the formula.

(c) Estimate the drag coefficient due to 'lock-in' by using a strip theory approach.

6.3 Galloping motions (instabilities) of a ship moored to a single point mooring system

Consider a ship moored to a single point mooring system (see Fig. 6.35). When the ship is in its equilibrium position, the incident current is in the longitudinal direction of the ship, i.e. along the positive x-axis. We will neglect the effects of wind and waves (in principle it is straightforward to

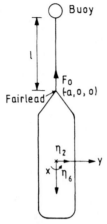

Fig. 6.35. Definition of parameters for analysis of stability of ships moored to single point mooring system.

include them in the analysis). The effect of the buoy motion will be neglected. According to Liapis (1985) this is a reasonable approximation.

We will study the stability of the sway and yaw motion of the ship. Liapis (1985) used the following equations for analysis of stability of sway and yaw motion

$$m_{yy} \frac{d^2\eta_2}{dt^2} + C_{yy}\eta_2 + C_{y\psi}\eta_6 = 0$$

$$C_{\psi y}\eta_2 + m_{\psi\psi} \frac{d^2\eta_6}{dt^2} + C_{\psi\psi}\eta_6 = 0$$

$$(6.73)$$

where

$$m_{yy} = M + A_{22}, \qquad m_{\psi\psi} = I_6 + A_{66}, \qquad C_{yy} = \frac{F_0}{l}$$

$$C_{y\psi} = -\left(\frac{a}{l} + 1\right)F_0 - \frac{\partial F_y}{\partial\psi}, \qquad C_{\psi y} = -\frac{a}{l}F_0,$$

$$C_{\psi\psi} = aF_0\left(\frac{a}{l} + 1\right) - \frac{\partial M}{\partial\psi}$$

Here η_2 is the sway motion of the centre of gravity, M is the mass, A_{22}

the added mass in sway, I_6 the mass inertia moment in yaw and A_{66} the added mass moment in yaw for the ship. $\partial M / \partial \psi$ is the rate of change of the current yaw moment on the ship with respect to the angle ψ at the equilibrium angle $\psi = 0$ when the ship is rotating in positive yaw angle direction. F_y is the transverse current force. $\partial F_y / \partial \psi$ can be explained in a similar way to $\partial M / \partial \psi$. F_0 is the tension in the bow hawser, l is the length of the bow hawser and $-a$ is the x-coordinate of the fairlead. The effect of the elasticity of the hawser is neglected. The restoring effect from the hawser that is incorporated in equation (6.73) can be shown by Fig. 6.36. From the figure we see that the variation in the angle $\Delta \psi$ on the bow hawser can be written as

$$\Delta \psi = (\eta_2 - a\eta_6)/l \tag{6.74}$$

The transverse force from the bow hawser on the ship can therefore be expressed as

$$-F_0(\Delta \psi - \eta_6) \tag{6.75}$$

Equations (6.74) and (6.75) explain the effect of the bow hawser in the C_{yy} and $C_{y\psi}$ term. By taking the yaw moment of the force expressed by (6.75), we obtain the bow hawser terms in $C_{\psi y}$ and $C_{\psi\psi}$.

    (a) What approximations are used in formulating equation (6.73)?

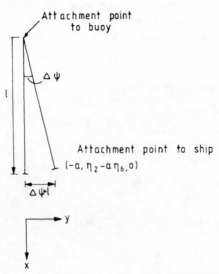

Fig. 6.36. Illustration used in analysis of restoring effect of a bow hawser on the sway and yaw motion of a ship.

(b) Express the bow hawser tension F_0 in terms of current loads on the ship and thrust from the propellers.

(c) Assume the time dependence of η_2 and η_6 can be written as $e^{\sigma t}$. Show that σ satisfies

$$a_4 \sigma^4 + a_2 \sigma^2 + a_0 = 0 \qquad (6.76)$$

where

$$a_4 = m_{yy} m_{\psi\psi}$$

$$a_2 = m_{yy} \left\{ aF_0 \left(\frac{a}{l} + 1 \right) - \frac{\partial M}{\partial \psi} \right\} + m_{\psi\psi} \frac{F_0}{l}$$

$$a_0 = -\frac{F_0}{l} \left(\frac{\partial M}{\partial \psi} + a \frac{\partial F_y}{\partial \psi} \right) > 0 \quad \text{(criterion for static stability)}$$

(d) The sway and yaw motions are stable if the real part of all solutions of σ are negative. According to Routh (1955) this means that $a_2 > 2(a_0 a_4)^{\frac{1}{2}}$. Show that this is equivalent to

$$(m_{yy} a^2 + m_{\psi\psi}) \frac{F_0}{l} - 2 \left[-m_{yy} m_{\psi\psi} \left(\frac{\partial M}{\partial \psi} + a \frac{\partial F_y}{\partial \psi} \right) \right]^{\frac{1}{2}} \left(\frac{F_0}{l} \right)^{\frac{1}{2}}$$

$$+ m_{yy} \left(aF_0 - \frac{\partial M}{\partial \psi} \right) > 0 \quad (6.77)$$

(e) Show that:
If the tension in the bow hawser $F_0 > F_{cr}$, the ship is stable for any length l of the bow hawser. If $F_0 < F_{cr}$, the ship is stable for $l < l_{cr}$ and unstable for $l > l_{cr}$.

Here

$$F_{cr} = \frac{a \dfrac{\partial M}{\partial \psi} - \dfrac{m_{\psi\psi}}{m_{yy}} \dfrac{\partial F_y}{\partial \psi}}{\dfrac{m_{\psi\psi}}{m_{yy}} + a^2} \qquad (6.78)$$

$$l_{cr} = F_0 \frac{\left(a^2 + \dfrac{m_{\psi\psi}}{m_{yy}} \right)^2}{\left\{ \left[-\dfrac{m_{\psi\psi}}{m_{yy}} \left(\dfrac{\partial M}{\partial \psi} + a \dfrac{\partial F_y}{\partial \psi} \right) \right]^{\frac{1}{2}} + \left[\dfrac{\Delta}{m_{yy}} \right]^{\frac{1}{2}} \right\}^2} \qquad (6.79)$$

where

$$\Delta = -a\left[m_{\psi\psi}\left(F_0 + \frac{\partial F_y}{\partial \psi}\right) + m_{yy}a\left(aF_0 - \frac{\partial M}{\partial \psi}\right)\right] \qquad (6.80)$$

(Hint: Consider the left hand side of equation (6.77) as a second degree polynomial with independent variable $t = (F_0/l)^{\frac{1}{2}}$ and study the sign of this polynomial.)

(f) Choose $a \approx L/2$, $m_{\psi\psi}/m_{yy} = (L/4)^2$ and set the propeller thrust equal to zero. (L = ship length.) Show that equation (6.79) can be written in the form

$$\frac{l_{cr}}{L} \approx \frac{C}{(\log_{10} \mathrm{Rn} - 2)^2} \frac{S}{A_{LS}} \qquad (6.81)$$

where S is the wetted area of the ship, $\mathrm{Rn} = U_c L/v$, $A_{LS} = LD$ and D is the draught. Use Figs. 6.11 and 6.13 to estimate typical values of $\partial M/\partial \psi$ and $\partial F_y/\partial \psi$ and give estimations of C in equation (6.81).

Calculate l_{cr} for a ship with $L = 300$ m, $S/A_{LS} = 4.0$ and when $U_c = 1.5$ knots. Choose a reasonable value of C.

(g) Discuss how propeller thrust and ship hull geometry can influence the stability.

7 VISCOUS WAVE LOADS AND DAMPING

Viscous flow phenomena are of importance in several problems related to wave loads on ships and offshore structures. Examples are wave loads on jackets, risers, tethers and pipelines, roll damping of ships and barges, slow-drift oscillation damping of moored structures in irregular sea and wind, anchor line damping, and 'springing' damping of TLPs.

The main factors influencing the flow are:

Reynolds number $Rn = UD/v$ (U = characteristic free stream velocity, D = characteristic length of the body, v = kinematic viscosity coefficient),

Roughness number $= k/D$ (k = characteristic cross-sectional dimension of the roughness on the body surface),

Keulegan–Carpenter number (KC) ($= U_M T/D$ for ambient oscillatory planar flow with velocity $U_M \sin((2\pi/T)t + \epsilon)$ past a fixed body),

Relative current number ($= U_c/U_M$ when the current velocity U_c is in the same direction as the oscillatory flow velocity $U_M \sin((2\pi/T)t + \epsilon)$),

Body form,

Free-surface effects,

Sea-floor effects,

Nature of ambient flow relative to the structure's orientation,

Reduced velocity ($U_R = U/(f_n D)$ for an elastically mounted cylinder with natural frequency f_n).

Sometimes $\beta = Rn/KC = D^2/(vT)$ is also used to characterize the flow. Detailed discussions of many of the factors mentioned above can be found in Sarpkaya & Isaacson (1981). For harmonically oscillating flow around a fixed circular cylinder of diameter D, we may write $KC = 2\pi A/D$, where A is the amplitude of oscillation of the fluid far away from the body. We then see that KC expresses the distance a free stream fluid particle moves relative to the body diameter.

MORISON'S EQUATION
Morison's equation (Morison *et al.* 1950) is often used to calculate wave loads on circular cylindrical structural members of fixed offshore

structures when viscous forces matter. Morison's equation tells us that the horizontal force dF on a strip of length dz of a vertical rigid circular cylinder (see Fig. 3.14) can be written as

$$dF = \rho \frac{\pi D^2}{4} dz\, C_M a_1 + \frac{\rho}{2} C_D D\, dz\, |u|\, u \qquad (7.1)$$

Positive force direction is in the wave propagation direction. ρ is the mass density of the water, D is the cylinder diameter, u and a_1 are the horizontal undisturbed fluid velocity and acceleration at the midpoint of the strip. The mass and drag coefficients C_M and C_D have to be empirically determined and are dependent on the parameters mentioned in the beginning of the chapter.

Considering deep water regular sinusoidal incident waves (see Table 2.1) and assuming C_M and C_D to be constant with depth (which might not be realistic), we may easily show that the mass-force decays with depth like $e^{2\pi z/\lambda}$. The drag force decays like $e^{4\pi z/\lambda}$ and is even more concentrated in the free-surface zone. When there is a wave node at the cylinder axis the mass-force will have a maximum absolute value and the drag-force will then be zero. The drag-force on a submerged strip will have a maximum absolute value when there is a wave crest or a wave trough at the cylinder axis. If viscous effects are negligible, it is possible to show analytically that Morison's equation is the correct asymptotic solution for large λ/D-values (see discussion of equation (3.34)). The C_M-value should then be two for a circular cross-section. If fluid acceleration can be neglected, Morison's equation is a reasonable empirical formulation for the time average force. Typical C_D- and C_M-values for transcritical flow past a smooth circular cylinder at $KC > \approx 40$ are 0.7 and 1.8. A roughness number k/D of 0.02 can represent more than 100% increase in C_D relative to the C_D-value for a smooth circular cylinder (Sarpkaya, 1985). This means the effect of roughness is more significant in oscillatory ambient flow than in steady incident flow.

The application of Morison's equation in the free-surface zone requires accurate estimates of the undisturbed velocity distribution under a wave crest. The prediction of the velocity distribution based on linear wave theory was discussed in connection with Fig. 2.2. A straightforward application of Morison's equation implies that the absolute value of the force per unit length is largest at the free-surface. This is unphysical since the pressure is constant on the free-surface. This means the force per unit length has to go to zero at the free-surface. It should also be noted that the position of the free-surface at the cylinder is affected by wave run up on the upstream side of the cylinder and a wave depression on the downstream side. The vertical position of the maximum absolute value of the local force has to be experimentally determined. The order of magnitude of the position may be 25% of the wave amplitude down in the fluid from the free-surface. We should note that Morison's equation cannot predict at all the oscillatory forces due to vortex shedding in the lift direction, i.e. forces orthogonal to the wave propagation direction and in the cross-sectional plane.

Morison's equation can be modified in the case of a moving circular cylinder. Consider a vertical cylinder and denote the horizontal rigid body motion of a strip of length dz by η_1. We can write the horizontal hydrodynamic force on the cylinder as

$$dF = \tfrac{1}{2}\rho C_D D \, dz(u - \dot{\eta}_1) \, |u - \dot{\eta}_1|$$
$$+ \rho C_M \frac{\pi D^2}{4} \, dz \, a_1 - \rho(C_M - 1) \frac{\pi D^2}{4} \, dz \, \ddot{\eta}_1 \qquad (7.2)$$

Dot stands for time derivative. Both u and a_1 are position-dependent. It should be noted that the inertia term does not depend on the relative acceleration term. The reason for this can be found by studying the analytical solution in the potential flow case. The Froude–Kriloff force (i.e. undisturbed wave pressure force) results in a horizontal force $\rho(\pi D^2/4) \, dz \, a_1$ which is independent of the rigid body motion (see chapter 3). The C_D- and C_M-values in equation (7.2) are not necessarily the same as in equation (7.1).

Morison's equation can also be applied to inclined members. To demonstrate this let us consider a cylinder inclined in a plane parallel to the wave propagation direction. The approach would be to decompose the undisturbed velocity and acceleration into components normal to the cylinder axis and components parallel to the cylinder axis, and then use Morison's equation with normal components of velocity and acceleration. The force direction will be normal to the cylinder axis. In the potential flow case, it can be proven that this is the correct expression. In the viscous case it means that we use the 'cross-flow' principle (see chapter 6). Actually what we are proposing to do for an inclined cylinder is not different from the vertical cylinder case. In the latter case there is also an undisturbed tangential velocity and acceleration component in the fluid. The effect of this was neglected in equation (7.1).

When the cylinder axis is not in the plane of the wave propagation direction, there exist different possibilities for formulations. Let us illustrate this using a submerged horizontal circular cylinder in waves where the wave propagation direction is orthogonal to the cylinder axis. A straightforward generalization of Morison's equation would be to write the horizontal and vertical force on a strip of length dy as

$$dF_1 = \rho \frac{\pi D^2}{4} \, dy \, C_M a_1 + \frac{\rho}{2} C_D D \, dy \, u(u^2 + w^2)^{\frac{1}{2}} \qquad (7.3)$$

$$dF_3 = \rho \frac{\pi D^2}{4} \, dy \, C_M a_3 + \frac{\rho}{2} C_D D \, dy \, w(u^2 + w^2)^{\frac{1}{2}} \qquad (7.4)$$

Here w and a_3 are vertical undisturbed fluid velocity and acceleration components at the midpoint of the strip. Chaplin (1988) has shown that improved correlation with experiments can be obtained if different C_M and C_D coefficients are assigned to the horizontal and vertical components.

When waves and currents are acting simultaneously, the combined effect should be considered. The normal approach is to add vectorially the wave-induced velocity and the current velocity in the velocity term of the Morison's equation. As mentioned in the beginning of the chapter, one should be aware that C_M- and C_D-values are also influenced by the presence of a current.

The design wave approach is often used in combination with Morison's equation. This means one analyses the load effect of a regular wave system. Often non-linear wave theories are used. A short term-statistical approach may also be applied (Vinje, 1980). However, more has to be learned about C_M- and C_D-values in this context. For instance, is it right to use just one C_M- and one C_D-value to represent a sea state? If so, is it then sufficient to use a characteristic Keulegan–Carpenter number, Reynolds number, roughness ratio etc. to estimate the C_M- and C_D-value? The answer is probably yes if the characteristic KC-number is large and the flow is not in a critical flow regime. What we mean by a large characteristic KC-number has to be precisely defined. In the case of extreme wave loads on a strip of a vertical circular cylinder close to the free surface $\pi H_{max}/D$ is a characteristic KC-number, where H_{max} is the most probable largest wave height. If $\pi H_{max}/D > \approx 40$ we could consider the KC-number to be large. A typical C_D-value for a smooth surface is then 0.7. It is of course easy to criticize Morison's equation. However, there is nothing better from a practical point of view. The reason for this is the very complicated flow picture that occurs for separated flow around marine structures.

Many attempts have been made to solve separated flow around marine structures numerically. Examples of methods used are

(a) Single vortex method (Brown & Michael, 1955; Faltinsen & Sortland, 1987)

(b) Vortex sheet model (Faltinsen & Pettersen, 1987)

(c) Discrete vortex method (Sarpkaya & Shoaff, 1979)

(d) Combination of Chorin's method and vortex-in-cell method (Chorin, 1973; Smith & Stansby, 1988)

(e) Navier–Stokes solvers (Lecointe & Piquet, 1985)

A general description of the state of the art is that the methods are generally limited to two-dimensional flow, that the methods have documented satisfactory agreement in some cases, but that they are presently not robust enough to be applied with confidence in completely new problems, where there is no guidance from experiments. Due to lack of proper turbulence modelling of wakes it is difficult to simulate the flow around a cylinder in the near-wake of another cylinder. Except for the single vortex method, the methods are often time consuming to apply to practical problems. One can end up spending a lot of computer time with numerical prediction methods for separated flow around offshore structures without gaining further knowledge. The single vortex method is limited to small KC-flow around bodies with sharp corners. The other methods have more general applicability.

An example of the simulated wake picture by the vortex sheet model of Faltinsen & Pettersen (1987) is presented in Fig. 7.1. The figure shows

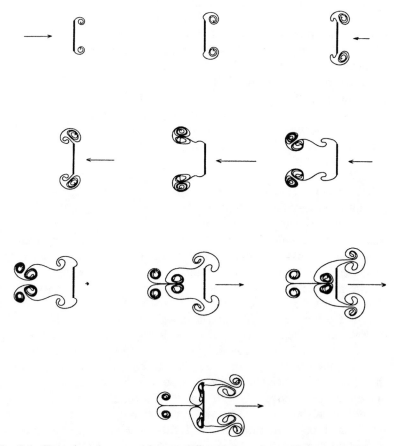

Fig. 7.1. Free shear layer positions at different times for oscillatory crossflow past a flat plate. $KC = U_M T/H = \pi$ (U_M = velocity amplitude of the ambient flow, T = oscillation period, H = half the height of the plate). $U_M t/H = 0.784, 1.584, 1.984, 2.384, 3.184, 3.984, 4.784,$ 5.584, 6.384, 6.984 (Faltinsen & Pettersen, 1987.)

oscillatory crossflow past a flat plate at different time steps. The arrow indicates the magnitude and direction of the instantaneous free-stream velocity. It seems that one pair of vortices return to the body in the last picture. This creates numerical problems. For larger amplitudes of oscillation of the ambient fluid, i.e. KC > 6–8, asymmetry develops in the wake. From flow visualizations it is seen that this causes vortices to pair and travel away from the plate.

In the following text we will base our discussion on viscous loads written in terms of Morison's equation. This means we will discuss how C_D and C_M depend on the parameters mentioned in the beginning of the chapter. In particular we will focus on the effect of body form and KC-number. For small KC-numbers the results have relevance for various damping problems. The results for large KC-numbers have relevance for wave loads on a jacket platform in extreme wave conditions. The structural members of a jacket have characteristic cross-dimensional diameters D of about 1 m. In a regular design wave with a return period of 100 years, the wave height H may be 30 m. This means the maximum KC-number in this case is $KC = \pi H / D = 94$.

Occurrence of flow separation strongly affects the drag forces. We will start out by discussing this and later on in the text discuss results for higher KC-numbers.

FLOW SEPARATION

When viscous effects matter for offshore structures, the flow will generally separate. A consequence of separation is that pressure forces due to viscous effects are more important than shear forces. There is some confusion about what is precisely meant by separation in unsteady flow. There is obviously no disagreement that the flow has separated if vortices can be clearly observed in the fluid, but it is difficult to set a precise limit on when one would call a flow separated or not separated. There is a need for this if approximate analytical and numerical methods are used for a description of the flow. Telionis (1981) has discussed criteria for flow separation in unsteady flow. One definition is: 'Unsteady separation occurs when a singularity develops in the classical boundary-layer equations'. Even if the boundary layer equations do not describe the physical fluid behaviour exactly around a separation point, one might argue that the occurrence of a singularity in the mathematical solution is an indication that the flow breaks strongly away from the body. On the other hand one might question if this is a sufficient description of separation in unsteady flow. The flow may also be called separated if it does not break strongly away from the body, for instance if there is a thin recirculating wake close to the body. A necessary requirement for a recirculating wake to be present in a two-dimensional flow is that the shear stress is zero at a point on the body surface, i.e. that the shear stress changes sign. This is the normal separation criteria for steady incident flow. For unsteady flow Telionis prefers to call the point of zero skin friction the 'detachment' point.

The flow around blunt-shaped marine structures with no sharp corners will not separate in oscillatory fluid motion at very low Keulegan–Carpenter numbers. On the other hand the flow around blunt-shaped marine structures will always separate in a steady current. In the case of combined incident current and oscillatory fluid motion around a blunt-shaped marine structure with no sharp corners it depends on the relative current number (see definition in the beginning of the section) if there exists a limiting KC-number for the flow to separate. Roughly speaking one may say there is a possibility that the flow will not separate if the ambient flow velocity in the current direction changes sign with time.

Sarpkaya (1986) has experimentally examined circular cylinders at low KC-numbers for different Reynolds numbers. No current was present. Sarpkaya reported separation occurring for KC-numbers as low as 1.25.

The fact that the flow separates does not mean that vortices or vortex shedding are easily observed. For instance, Bearman (1985) states in a paper, previous to Sarpkaya (1986), that nobody seems to have observed vortices at KC-values less than three. However, observation of vortices does not necessarily mean that vortices are convected away from the vicinity of the body, i.e. that vortex shedding occurs. For instance, Bearman et al. (1981) studied a flow around a circular cylinder at KC = 4 where a pair of vortices generated during one half cycle did not survive into the next half cycle. When the flow reversed and the vortices were convected back to the cylinder, their circulations appeared to be cancelled by vorticity of opposite sign in the boundary layer of the cylinder. A systematic study of when vortex shedding starts has not been attempted, but as an example, Williamson (1985) reported symmetric vortex shedding occurring for a KC as low as four. By comparing this case to Bearman et al.'s (1981) studies, it is obvious that it is not only KC that determines when vortex shedding starts: for instance the Reynolds number must also be an important parameter.

By analysing high Reynolds number laminar start-up flow around bodies, we can determine in a qualitative way when flow separation occurs for subcritical flows.

Consider a two-dimensional stationary body in a high Reynolds number flow. Outside the body the fluid domain extends to infinity. The boundary layer on the body surface is assumed to be thin and laminar. The ambient flow velocity U_∞ is written as

$$U_\infty = U_M \sin(\omega t - \alpha) + U_c \qquad (7.5)$$

where U_M, U_c are constants, t is the time variable, $\omega = 2\pi/T$ is the circular frequency of oscillation and α is a phase angle. We will select α so that U_∞ is zero at $t = 0$. This means $\alpha = \sin^{-1}(U_c/U_M)$. Initially, the vorticity is zero. We will assume that $U_c/U_M < 1$. This means the ambient flow velocity changes direction with time.

The flow will be analysed until the ambient flow velocity reaches its first maximum value, i.e. until $\omega t = \pi/2 + \alpha$ (see Fig. 7.2). At that time the free stream acceleration is zero. We will investigate for which KC-value and U_c/U_M-ratios the flow separates during this initial time period. The separation criterion to be used is that the shear stress vanishes at a point on the body surface. This separation criterion is for purpose of estimation. It is not generally accepted for unsteady flow. If the flow has not separated when $\omega t = \pi/2 + \alpha$ i.e. when the force according to Morison's equation (see equation (7.1)) is only due to drag forces, it is assumed that the flow separation has no influence on the drag forces. This idea has no meaning if the ambient flow never changes direction with time. The model is expected to behave better the smaller the ratio U_c/U_M is.

Consider a coordinate system (x, y) where x and y are locally tangential and normal to the body surface. The tangential fluid velocity u in the boundary layer is approximated by

$$u(x, y, t) = u_0(x, y, t) + u_1(x, y, t) + \ldots \tag{7.6}$$

where u_0 and u_1 are first- and second-order approximations to u. From Schlichting (1979: p. 410) it follows that the boundary layer equations can be approximated by

$$\frac{\partial u_0}{\partial t} - \nu \frac{\partial^2 u_0}{\partial y^2} = \frac{\partial U(x, t)}{\partial t} \tag{7.7}$$

where $U(x, t)$ is the tangential fluid velocity just outside the boundary layer. The boundary conditions are

$$u_0 = 0 \text{ at } y = 0 \text{ and } u_0 = U(x, t) \text{ at } y = \infty$$

Further

$$\frac{\partial u_1}{\partial t} - \nu \frac{\partial^2 u_1}{\partial y^2} = U \frac{\partial U}{\partial x} - u_0 \frac{\partial u_0}{\partial x} - v_0 \frac{\partial u_0}{\partial y} \tag{7.8}$$

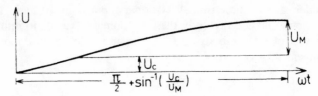

Fig. 7.2. Ambient flow velocity used in start-up flow studies (see equation (7.5)).

with boundary conditions

$$u_1 = 0 \text{ at } y = 0, \qquad \frac{\partial u_1}{\partial y} = 0 \text{ at } y = \infty$$

In equation (7.8) v_0 is the first-order approximation of the vertical velocity in the boundary layer. When u_0 is known, it follows from the continuity equation and body boundary conditions that

$$v_0 = - \int_0^y \frac{\partial u_0}{\partial x} \, dy \tag{7.9}$$

By solving a potential flow problem outside the boundary layer we can determine $U(x, t)$. We write

$$U(x, t) = (U_M \sin(\omega t - \alpha) + U_c)\Phi(x) \tag{7.10}$$

The approximate solution (equation (7.6)) has validity only during an initial phase when diffusion effects expressed by (equation (7.7)) dominate over convection effects. However, it is difficult to know a priori how large the values of ωt can be for the solution to be valid.

Equation (7.7) with boundary conditions can be solved analytically (Faltinsen (unpublished)). We can formally write the shear stress on the body as

$$\left.\frac{\partial u}{\partial y}\right|_{y=0} = \sqrt{\left(\frac{\omega}{\nu}\right)} U_M \Phi(x) \left[f_1 + \frac{U_M}{\omega} \frac{d\Phi}{dx} f_2 \right] \tag{7.11}$$

where f_1 and f_2 are independent of the body shape, but functions of U_c/U_M, α and ωt. Numerical results when $U_c/U_M = 0$ and $\alpha = 0$ in the form of $0.5\pi f_1/f_2$ as functions of ωt are presented in Fig. 7.3. When $\omega t \rightarrow 0$, we see that $f_1/f_2 \rightarrow \infty$.

It follows from equation (7.11) and the signs of f_1 and f_2, that separation (i.e. $\partial u/\partial y\,|_{y=0} = 0$) can only occur when $d\Phi/dx$ is negative, which fits with our intuition about adverse pressure gradients. Further, initial separation will occur at the point of minimum $d\Phi/dx$, which may or may not occur at the downstream stagnation point of the body. For bodies with a sharp corner $d\Phi/dx$ is infinite. From equation (7.11) it follows that separation occurs immediately at sharp corners.

Let us study a circular cylinder. This means

$$\Phi(x) = 2 \sin \frac{x}{R}$$

with $x = 0$ and $x = \pi R$ corresponding to the forward and aft stagnation points on the cylinder. The separation condition for a circular cylinder can then be written as

$$\frac{\pi}{2} f_1 + f_2 \, KC \cos \frac{x}{R} = 0$$

This means that the separation angles $\pm |\theta_s|$ are found from

$$\theta_s = \frac{x_r}{R} = \cos^{-1}\left[-\frac{f_1}{f_2}\frac{\pi}{2\mathrm{KC}} \right] \tag{7.12}$$

From this it follows that separation starts at the aft stagnation point when

$$\frac{f_1}{f_2}\frac{\pi}{2} = \mathrm{KC} \tag{7.13}$$

The left-hand side of (7.13) is a function of ωt and U_c/U_M. This function has been plotted in Fig. 7.3 for $U_C/U_M = 0$. Equation (7.13) tells us directly at what time a flow with a given KC-value and U_c/U_M ratio will start to separate. Equation (7.12) can be used to find the separation angles as a function of time. Results from a numerical example are shown in Fig. 7.4 for $\mathrm{KC} = 3.1$ and $U_c/U_M = 0$. We note that the separation starts at time $\omega t = 1.045$. This can also be seen from Fig. 7.3. The limiting KC-values $\mathrm{KC}_{\mathrm{lim}}$ for separation to occur, will be

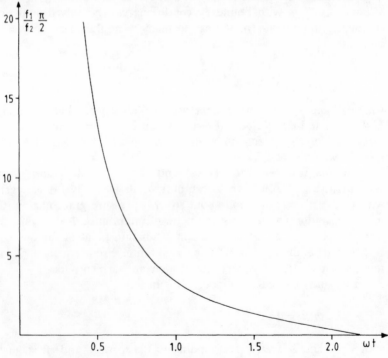

Fig. 7.3. Calculation of the laminar boundary layer functions f_1 and f_2 (see equation (7.11)) for sinusoidal ambient velocity start-up flow. The data are presented as a function of non-dimensionalized time ωt. The current velocity is zero.

the KC-value where separation starts at $\omega t = \pi/2 + \alpha$. When $U_c/U_M = 0$ this will be the value of $0.5\pi f_1/f_2$ at $\omega t = \pi/2$ (see equation (7.13) and Fig. 7.3). Thus KC_{lim} equals 1.2.

This procedure can be generalized to axisymmetric flow. Values for KC_{lim} are presented in Fig. 7.5 for a circle and a sphere as a function of U_c/U_M. The results presented in Fig. 7.5 do not depend on Reynolds number. The reason is that the separation point determination expressed by (7.12) is Reynolds number independent. This is based on the assumption of laminar, high Reynolds number start up flow.

In the case of a sphere we may compare the KC-values for which separation occurs with the experimental values observed by Zhao et al. (1988). They investigated a hemisphere of diameter 1 m. In calm water the centre of the hemisphere was at the still water free-surface level. Visualization tests were performed to study the occurrence of flow separation around the hemisphere in combined current and regular waves. The model was restrained from oscillating. Only the flow at the free-surface was examined by placing confetti with a diameter of 3 mm on the water surface and taking photographs. When the separated region is very close to the body the confetti technique may not be effective

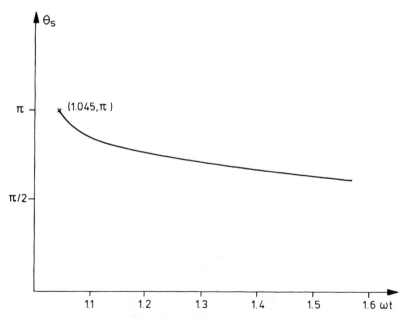

Fig. 7.4. Time development of separation angle θ_s around a circular cylinder. ($\theta = 0$ corresponds to the upstream stagnation point on the cylinder.) The ambient flow velocity is sinusoidal and zero at $\omega t = 0$, KC = 3.1. $U_c/U_M = 0$. (KC and U_c/U_M defined in the beginning of the chapter.)

enough to detect the separated flow. There was no way to confirm that the flow always stayed laminar in the boundary layer. The current was simulated by towing the model against the wave propagation direction. The results are presented in Fig. 7.6. The free-surface effect and the circular motion of the ambient flow makes the experimental conditions different from our theoretical assumptions. On the other hand, the free-surface wave parameter $\omega(D/g)^{\frac{1}{2}}$ (g = acceleration of gravity, D = diameter) did not seem to influence the observations significantly. Furthermore, circular ambient fluid motion may not be very different from planar ambient fluid motion when the wake is prevented from rotating around the body. This seems to be the experience from vertical cylinders in waves, while submerged horizontal cylinders in waves will behave quite differently in planar ambient flow and in circular ambient flow (Chaplin, 1984).

In a qualitative sense, one may say that the theoretical results show good agreement with experiments; but obviously one should be critical of the theoretical predictions close to $U_c/U_M = 1.0$.

OSCILLATORY NON-SEPARATED FLOW

We will discuss how to predict forces on bodies in non-separated flow. In order to evaluate the shear forces for laminar flow we will use a solution found in many textbooks on fluid dynamics (see for instance Schlichting, 1979: pp. 428–9). We will assume that the outer flow U_e outside the boundary layer can be written as

$$U_e(x, t) = U_0(x) \cos \omega t \qquad (7.14)$$

It is assumed that the oscillatory amplitudes are small. This means we may neglect quadratic terms in the boundary layer equations. For

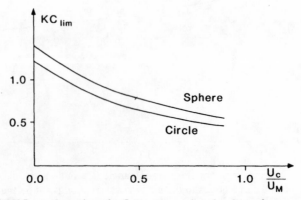

Fig. 7.5. KC-number when the flow separated at the time of zero ambient flow acceleration following a sinusoidal start-up flow. (KC and U_C/U_M are defined at the beginning of the chapter.)

laminar flow we can write

$$\frac{\partial u}{\partial t} - \nu \frac{\partial^2 u}{\partial y^2} = \frac{\partial U_e}{\partial t} \tag{7.15}$$

A steady-state solution of this equation with the boundary conditions $u = 0$ at $y = 0$ and $U = U_e$ at $y \to \infty$ is

$$u = U_0(x)[\cos \omega t - e^{-\eta} \cos(\omega t - \eta)] \tag{7.16}$$

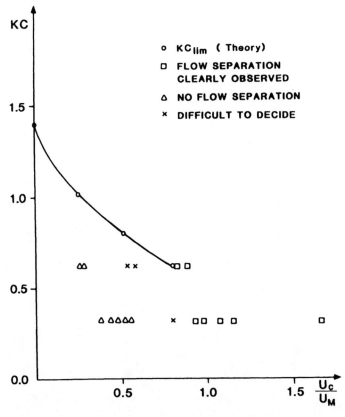

Fig. 7.6. Comparison between theory and experimentally observed flow separation around a hemisphere in regular waves and current. The centre of the hemisphere is in the mean free-surface. A vertical cylinder is fitted to the hemisphere above the free-surface to avoid sharp corners on the body. ($KC = U_M T/D$, U_M = maximum velocity of ambient oscillatory flow, T = oscillation period, D = diameter of the sphere, U_c = current velocity in the same direction as incident oscillatory velocity.) (Zhao et al., 1988).

Here

$$\eta = \left(\frac{\omega}{2\nu}\right)^{\frac{1}{2}} y \tag{7.17}$$

The velocity distribution is shown in Fig. 7.7 for different time instants. The velocity does not have the same phase at all y-coordinate points in the boundary layer. Equation (7.16) can be used to find the boundary layer thickness. There are many different definitions of boundary layer thickness. However, let us consider the point where $U_0(x)e^{-(\omega/(2\nu))^{\frac{1}{2}}y} = 0.01U_0(x)$. This corresponds to $(\omega/(2\nu))^{\frac{1}{2}}y = 4.6$. For $\nu = 10^{-6}\,\mathrm{m^2\,s^{-1}}$ and $T = 10\,\mathrm{s}$ this means $y = 0.008\,\mathrm{m}$.

The wall skin friction force (shear force) per unit area can be written as

$$\tau_{\mathrm{w}} = \mu\left(\frac{\partial u}{\partial y}\right)_{y=0} = \mu U_0\left(\frac{\omega}{\nu}\right)^{\frac{1}{2}} \cos\left(\omega t + \frac{\pi}{4}\right) \tag{7.18}$$

which osciilates with a phase lead of 45° relative to the external stream.

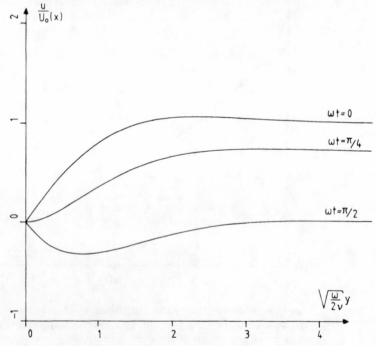

Fig. 7.7. Example of the horizontal (u) velocity distribution in a laminar boundary layer along a flat plate for harmonically oscillating outer flow (equation (7.16)). (y = coordinate normal to the flat plate, $y = 0$ is at the flat plate, ω = circular frequency of oscillation, $U_0(x)$ = velocity amplitude just outside the boundary layer.)

Let us apply this solution to non-separated oscillatory laminar flow past a circular cylinder. The free stream velocity far away from the body will be written as $U_\infty \cos \omega t$. The solution U_e outside the boundary layer can then be written as

$$2U_\infty \sin \theta \cos \omega t$$

From equation (7.18) the shear stress can be written as

$$\tau_w = \mu 2U_\infty \left(\frac{\omega}{\nu}\right)^{\frac{1}{2}} \sin \theta \cos\left(\omega t + \frac{\pi}{4}\right)$$

The resulting force on the cylinder in the incident flow direction can be expressed by

$$F_1 = \int_0^{2\pi} \tau_w \sin \theta \, R \, d\theta$$

$$= \mu U_\infty \left(\frac{\omega}{\nu}\right)^{\frac{1}{2}} \cos\left(\omega t + \frac{\pi}{4}\right) D\pi \qquad (7.19)$$

By writing

$$\cos\left(\omega t + \frac{\pi}{4}\right) = \frac{\sqrt{2}}{2} \cos \omega t - \frac{\sqrt{2}}{2} \sin \omega t$$

we see that there is one force component in phase with the velocity and another force component in phase with the acceleration. It should be noted that the force is linear with respect to U_0. The drag-force is normally said to be proportional to the square of the velocity. If we calculate the drag coefficient, we find

$$C_D{}^F = \frac{\mu U_\infty \left(\frac{\omega}{\nu}\right)^{\frac{1}{2}} D\pi / \sqrt{2}}{\frac{\rho}{2} U_\infty^2 D} = 2\pi \left(\frac{\pi}{RnKC}\right)^{\frac{1}{2}} \qquad (7.20)$$

However this is not the total viscous drag force. There is also an influence of viscosity on the normal stresses (Batchelor, 1970: p. 355). According to Stokes (1851) this results in a total drag force on a circular cylinder that is twice as large as in equation (7.20). Wang's (1968) analysis shows that we may write the drag coefficient on a circular cylinder in unseparated laminar high Reynolds number flow as

$$C_D = \frac{3\pi^3}{2KC} [(\pi\beta)^{-\frac{1}{2}} + (\pi\beta)^{-1} - \tfrac{1}{4}(\pi\beta)^{-\frac{3}{2}}] \qquad (7.21)$$

where $\beta = Rn/KC = D^2/(\nu T)$. This differs both from equation (7.20) and Stokes' solution. The drag coefficient can be very large, (actually it goes to infinity when $KC \rightarrow 0$), but this does not mean that the drag-force is large.

Johnson (1978) has given empirical formulas for shear stress that apply to turbulent flow along fixed plane surfaces. Outside the boundary layer the flow is oscillating harmonically. When the surface is smooth, Johnson writes the maximum wall shear stress τ_{wm} as

$$\frac{|2\tau_{wm}|}{\rho U_{lm}^2} = 0.09 \, RE^{-0.2} \tag{7.22}$$

where

$$RE = \frac{U_{lm}^2}{\omega \nu}, \tag{7.23}$$

Here U_{lm} is maximum tangential velocity just outside the boundary layer. $RE = 10^5$ is proposed as an engineering criterion for transition to turbulence. Equation (7.22) can be used as a basis for establishing slow-drift viscous surge damping of a ship. We can multiply $|\tau_w|$ by the wetted body surface and $(1 + k)$, where k is a form factor with a similar meaning as for viscous resistance of a ship (see discussion after equation (6.25)). Typical k-values for viscous resistance vary between 0 and 0.5.

SEPARATED FLOW AT SMALL KC-NUMBERS

Bearman (1985) has reported the behaviour of the flow around a circular cylinder for different KC-numbers when the ambient flow is planar and harmonically oscillating. For KC < 7 he says the flow is symmetric. This implies zero lift force. However, we cannot set up precise limits for the onset of asymmetric flow. For instance, the data by Bearman (1988) show that KC = $\approx$5 is a limiting value between symmetric and asymmetric flow. Bearman (1985) reports that a change in the vortex shedding appears at KC = 7 or 8. The majority of the vortex shedding takes place only on one side of cylinder. The flow has a strong memory in this range. Between KC = 15 and 25 a vortex is shed before the end of a half cycle. In addition a second vortex forms from the same shear layer. Vortices shed previously combine with the new vortices so that vortex pairs travel away in two trails at roughly 45° to the direction of the main flow. The directions can be switched. This depends on the starting condition. At KC > 25 at least three full vortices are shed per half cycle. The wake resembles the wake in steady flow, i.e. the vortices tend to form a street behind the body.

Results for low KC-numbers (KC < $\approx$10) have relevance for damping of slow drift motions of moored structures and roll damping of ships and barges. Graham (1980) has done an interesting analysis for KC → 0, which can be used to explain experimental results for KC < $\approx$10. His analysis includes only the effect of separated flow on the pressure distribution around the body. Viscous shear forces have to be added. Graham argued that the different behaviour of the drag coefficients of different sections at small values of KC is due to the relative strengths of the vortex shedding, a circular cylinder obviously being much weaker than a flat plate, for example. He assumes that the vortex flow for a small Keulegan–Carpenter number depends only on the local flow around each sharp edge. The edge is characterized by its internal angle δ. That means for a flat plate $\delta = 0$ and for a square section $\delta = \pi/2$. The vortex force F_V on a sharp edge acts along the perpendicular to the bisector of the edge angle. It can be shown to be proportional to KC^η, where $\eta = (2\delta - \pi)/(3\pi - 2\delta)$.

This means that, for example, the drag coefficient C_D for small KC-numbers should vary as

$$C_D \propto KC^{-\frac{1}{3}} \text{ for a flat plate}$$
$$\propto \text{constant for a square section}$$
$$\propto KC \text{ for a circular cylinder (regarded loosely as a}$$
$$\text{sharp-edged section with } \delta = \pi)$$

These variations are in quite good agreement with measured data for $KC < 10$, for bodies in planar ambient flow (Graham, 1980). The experimental data for a diagonal and facing square have later been corrected by Bearman et al. (1984a). The experimental results show that

$$\begin{aligned}
C_D &\simeq 8.0 \, KC^{-\frac{1}{3}} \quad &\text{(flat plate)} \\
C_D &\simeq 5.0 \quad &\text{(diagonal square)} \\
C_D &\simeq 3.0 \quad &\text{(facing square)} \\
C_D &\simeq 0.2 \, KC \quad &\text{(circular cylinder)}
\end{aligned} \tag{7.24}$$

The results for a circular cylinder are only of a qualitative nature. This is evident by comparing with Sarpkaya's (1986) experimental results. Sarpkaya examined a circular cylinder in planar oscillatory flow of small amplitude. The results were presented for different KC and β-values ($\beta = Rn/KC$). When the boundary layer flow was laminar and the flow did not separate (i.e. $KC < \approx 1$) Wang's formula (see equation (7.21)) agreed well with the experimental results (see Fig. 7.8). When $KC > \approx 2$ we note that the C_D-values start to increase with KC-number. This is what Graham's asymptotic solution says. However, one should note that his theory is based on the separation point being fixed. This is not the case for separated flow around a circular cylinder, particularly at small KC-numbers.

The nature of the ambient flow is important. The results by Chaplin (1984, 1988) demonstrate this. Chaplin (1988) presented data for a submerged horizontal circular cylinder in beam sea waves at small KC-numbers. The ambient flow was either nearly circular or elliptical. Chaplin argued that a circulation is set up around the cylinder. This has a large influence on the mass coefficients. In the particular case of a deeply submerged cylinder in nearly circular orbital flow he found that

$$C_M \approx 2 - 0.21\, KC^2 \tag{7.25}$$

The KC-number was less than two and the Reynolds number was of the order of 10^4. Equation (7.25) shows a very strong influence of KC on the C_M-value. This is opposite to what happens to the C_M-value in planar oscillatory flow for small KC-numbers.

Eddy-making roll damping

The results by Graham (see equation (7.24)) demonstrate the different degrees of damping due to vortex shedding when a body makes small oscillations relative to the fluid. If we apply his results to eddy-making roll damping of a rectangular cross-section, it means that the damping is independent of the KC-number. Another example is roll damping due to bilge keels. We then start out by writing the normal force F_n on the bilge keel as

$$F_n = -\frac{\rho}{2}\, C_D bl\, |u|\, u \tag{7.26}$$

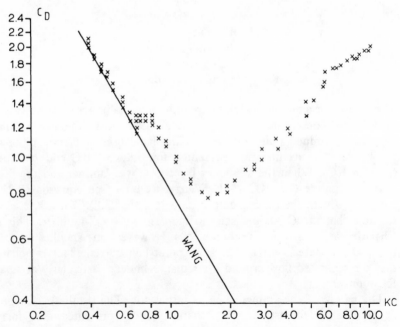

Fig. 7.8. Drag coefficients C_D versus Keulegan–Carpenter number KC for a smooth circular cylinder at $\beta = 1035$. ($\beta = Rn/KC = D^2/(vT)$, D = cylinder diameter, T = period of oscillation. $\times$ Experiments (Sarpkaya) —Theory (Wang). (Adapted from Sarpkaya, 1986).

where b and l are respectively the breadth and length of the bilge keel, and u is the relative velocity normal to the bilge keel as if the bilge keel was not there. For instance, if the bilge keel were fitted to a circular cylinder of radius R and the cylinder were rolling about its axis we can write $u \approx R\dot{\eta}_4$ when no incident waves are present. If we neglect the effect of the curvature in the vicinity of where the bilge keel is fitted to the hull, we can approximate the local flow around the bilge keel with the flow around a flat plate of breadth $2b$ in transverse oscillatory flow. According to Graham's formula $C_D = 8.0\,(KC)^{-\frac{1}{3}}$ for harmonically oscillating ambient flow. We can write $KC = u_{max}\,T/(2b)$. The separated flow around the bilge keel will also affect the pressure distribution on the hull. However, for a circular cylinder the hull pressure does not cause a roll moment about the cylinder axis. By taking the roll moment of equation (7.26) about the roll axis we will find the following roll damping moment for a circular cylinder

$$F_4 = -4\rho(KC)^{-\frac{1}{3}}blR^3 \frac{d\eta_4}{dt} \left| \frac{d\eta_4}{dt} \right| \qquad (7.27)$$

This shows that bilge keel damping is strongly influenced by the KC-number. For a general body shape, for instance a typical midship section of a ship, the approach outlined is only of a qualitative nature. The undisturbed velocity u will for instance be significantly larger when the flow accelerates around the bilge radius of a typical midship section. Empirical formulas for the bilge keel damping can be found in Kato (1966) and Ikeda et al. (1977b).

Viscous slow-drift damping of moored structures

The eddy-making damping is important for analysis of slow-drift motions of moored structures. One exception is surge motions of a ship where equations (7.18) or (7.22) can be used to estimate viscous damping.

The eddy-making damping for sway and yaw motions of a ship can in some cases be calculated by strip theory and the 'cross-flow principle' (see discussion of Fig. 6.11). This means that we write the damping force similar to that in equation (6.26) and analyse cross-flow past two-dimensional cross-sections of the ship.

The drag coefficients depend for instance on free-surface effects, beam–draught ratio, bilge keel dimensions, bilge radius, current, Reynolds number, roughness ratio and Keulegan–Carpenter (KC) number. Some of these effects will be discussed below. In chapter 6 we discussed $KC = \infty$ more extensively.

Free-surface effects

The free-surface acts similar to an infinitely long splitter plate (see discussion in chapter 6 on free-surface effects on C_D-values for ship sections). This means that one can apply drag coefficients for the double body with splitter plates. The doubly body consists of the submerged body and the image body above the free-surface. The splitter plate effect causes a clear lowering effect on the drag coefficient for high KC-numbers. When the KC-number is low the eddies will stay symmetric for the double body without a splitter plate which means the free-surface has little effect for low KC-numbers.

Beam–draught ratio effects
In general there is little dependence on beam–draught ratio for rectangular cross-sections. One exception may be for beam–draught ratios lower than ≈ 1 and KC smaller than ≈ 10 (see Fig. 7.9).

Bilge keel effects
The drag coefficients for low KC-numbers depend significantly on the bilge keel. This is illustrated in Fig. 7.10. We note that the effect is strongest for the lowest KC-values. For large KC-values the drag coefficient is not very sensitive to the breadth of the bilge keel.

Bilge radius effects
There is a strong effect of the bilge radius on the drag coefficients. Increasing the bilge radius means decreasing the drag coefficient (see Fig. 7.11).

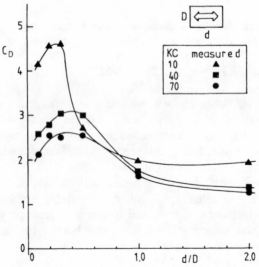

Fig. 7.9. Drag coefficient C_D as a function of aspect ratio d/D. Rectangular cylinders with sharp corners. D = Projected width. (KC = $U_M T/\dot{D}$, U_M = forced harmonic velocity amplitude of the cylinder, T = oscillation period.) (Tanaka *et al.*, 1982.)

Effect of laminar or turbulent flow (scale effects)

When separation occurs at sharp corners like bilge keels, there are not any severe scale effects. For high KC-number-flow around midship sections with no sharp corners, the drag coefficient in model scale may be roughly speaking twice the value in full scale. The scale effect on the drag coefficient at small KC-numbers is not so severe. The latter case is relevant for slow-drift oscillations in the absence of current.

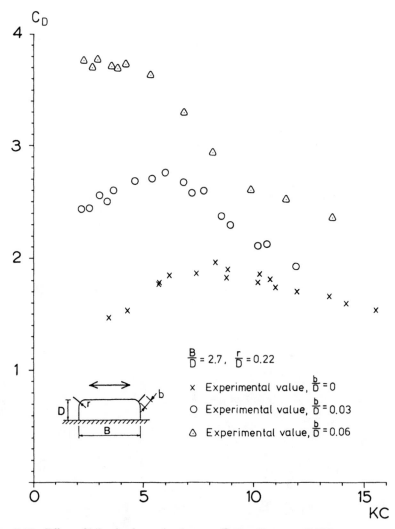

Fig. 7.10. Effect of bilge keels on the drag coefficient C_D at small KC-numbers. ($KC = U_M T/D$, U_M = ambient oscillatory flow velocity amplitude in the beam direction, T = oscillation period.) (Experimental results by Faltinsen & Sortland, 1987.)

The discussion above for eddy-making damping of ship sections is summarized in Table 7.1.

The discussion of eddy-making damping of slow-drift sway and yaw motion of a ship also has relevance for slow-drift damping of other structures, for example a TLP.

SEPARATED FLOW AT HIGH KC-NUMBERS

Knowledge of drag coefficients at high KC-numbers is important in predicting wave loads on jackets and risers in extreme weather conditions.

Table 7.1

Effect of:	KC < ≈10	KC > ≈10
Free surface	no	yes
Beam/draught	no	yes/no
Bilge keel	yes	yes
Bilge radius	yes	yes
Scale effect:		
(a) No sharp corners	yes/no	yes
(b) Bilge keels	no	no

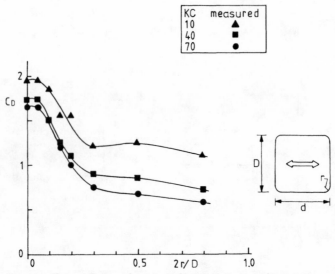

Fig. 7.11. Drag coefficient C_D as a function of bilge radius r ($d/D = 1.0$). (KC = $U_M T/D$, U_M = forced harmonic velocity amplitude of the cylinder, T = oscillation period.) (Tanaka *et al.*, 1982.)

Bearman *et al.* (1979) presented measurements of a series of two-dimensional bodies in plane oscillatory flow for KC-values between 3 and 70. They concentrated on bluff bodies with sharp-edged separation and measured the in-line force on a flat plate, square, diamond and circular cylinder. Beyond a KC-value of about 10 to 15 the curves for the flat plate, circular and diamond cylinders are remarkably similar, and by KC = 50 the C_D-values are all only a little higher than their steady flow values, C_D^∞. To demonstrate this similarity, the values of C_D above KC = 10 are shown plotted in Fig. 7.12 where they have been divided by their steady flow value. The square section cylinder showed a different trend.

We will try to explain the results in Fig. 7.12 by estimating the increased effective incident flow due to returning eddies. In order to estimate the drag coefficient we will study the force when the ambient velocity

$$U_\infty = U_M \sin \omega t \qquad (7.28)$$

is a maximum or a minimum, i.e. at $\omega t = ((2n + 1)/2)\pi$, $n = 0, 1, \ldots$ To illustrate the procedure we will study the force on the body when $\omega t = 3\pi/2$. The eddies that for instance are created at $\omega t = \pi/2$ then return as an increased incident flow.

The circulation Γ of the eddies will be estimated by a quasi-steady approach. Schmidt & Tilman (1972) measured the circulation of eddies as a function of longitudinal distance x from a circular cylinder in

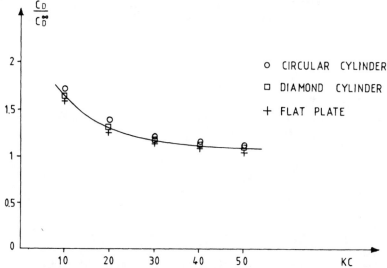

Fig. 7.12. The ratio between C_D at a given KC and C_D^∞ at KC = ∞ (i.e. steady incident flow). (Bearman *et al.*, 1979.)

incident flow with ambient velocity U_∞ along the x-axis (see Fig. 7.13). According to this figure the maximum circulation is

$$\Gamma \approx 5U_\infty R$$

where R is the cylinder radius.

When the eddies have obtained their maximum circulation, their strength decreases with distance (see Fig. 7.13). In the quasi-steady analysis of the oscillatory flow it is important to account for this effect. It means that the effective incident flow is a function of the distance the eddies have been convected. In order to quantify this effect we will use empirical circulation reduction formulas that are often used in numerical studies by the discrete vortex method. We will assume $\Gamma = \Gamma_0 e^{-\alpha s/R}$ where Γ_0 is the initial strength of a vortex and s the distance the vortex has been convected. Different values of α have been used in numerical studies. We will set $\alpha = 0.1$ in our example. However, this represents a larger decay factor than Fig. 7.13 shows and demonstrates sources of inaccuracies in our simple model. Since $KC = UT/D$ is proportional to the distances the shed vortices are convected, we may translate s/R into KC. For the eddies that are created at $\omega t = \pi/2$ and are returning to the body at $\omega t = 3\pi/2$ we can then write $\Gamma = \Gamma_0 e^{-0.1(2KC/\pi)}$. This means that

$$\Gamma = 5U_M R e^{-0.2(KC/\pi)} \tag{7.29}$$

where we have set $U_\infty = U_M$, i.e. the velocity of the ambient flow at $\omega t = \pi/2$. This is valid for a circular cylinder.

Due to the results in Fig. 7.12 we will assume equation (7.29) has validity for separated flow past any two-dimensional body shape. We will try to quantify the increased incident flow due to the returning eddies.

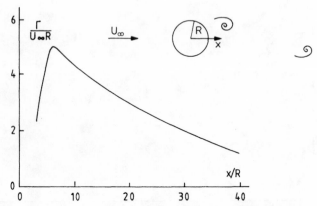

Fig. 7.13. Vortex strength versus distance for steady incident flow past a circular cylinder (Γ = circulation of a vortex). (Adapted from Schmidt & Tilman, 1972.)

This will be done by studying the effect of two vortices, shed from alternate sides at $\omega t \approx \pi/2$ (see Fig. 7.14). The inflow due to the vortices will vary with time and be dependent on the position relative to the cylinder. By inflow we mean the incident velocity when the body is not there. This is the reason why we have dotted the cylinder in the lower part of the figure. In order to illustrate that we are using a simplified model we have drawn the returning vorticity as two discrete vortices. The average velocity at the cylinder centre can be estimated from

$$\bar{u} = \frac{2}{l} \int_0^{l/2} d\xi \frac{\Gamma R}{2\pi} \left\{ \frac{1}{\xi^2 + R^2} + \frac{1}{\left(\frac{l}{2} - \xi\right)^2 + R^2} \right\} \qquad (7.30)$$

where we have used the potential flow solution for the velocity induced

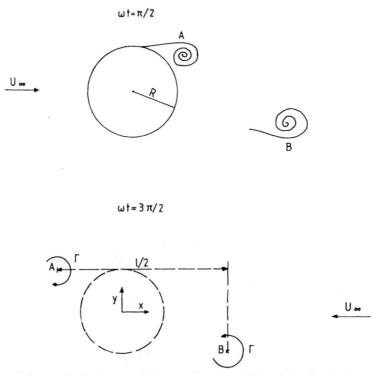

Fig. 7.14. Illustration to show that two returning vortices created half a period previously represent an increase in effective incident flow at present time. ($\pm\Gamma$ = circulations of vortices, ω = circular frequency of oscillation.)

by a vortex and $l/2$ is the horizontal distance between the vortices. Equation (7.30) can be integrated analytically. We find

$$\bar{u} = \frac{2\Gamma}{\pi l} \tan^{-1}\left(\frac{l}{2R}\right)$$

According to von Karman's analysis $2R/l = 0.281$ (see equation (6.16)). This means

$$\bar{u} = 0.83\, \Gamma/l = 0.58 U_M\, e^{-0.2(KC/\pi)} \tag{7.31}$$

The effective incident flow to the cylinder is $U_M + \bar{u}$. We will assume that the effect of eddies generated in previous cycles is negligible. This means

$$\frac{C_D}{C_D{}^\infty} \sim \frac{(U_M + \bar{u})^2}{U_M{}^2} = (1 + 0.58e^{-0.064KC})^2 \tag{7.32}$$

This is in good agreement with the results in Fig. 7.12. It even agrees with the experimental values at $KC = 10$. Due to the quasi-static assumptions one should not expect agreement for KC-values below 25. One should be careful in applying equation (7.32) to new situations since several major assumptions were made, for example equation (7.29) for the reduction of vortex circulation.

Lift forces

For large KC-numbers Verley (1982) and Bearman et al. (1984b) proposed a quasi-steady model to calculate the vortex-induced transverse force on a unit length of a fixed circular cylinder in oscillatory flow. They used the force in steady flow as a basis, i.e.

$$F_L(t) = \tfrac{1}{2}\rho U_\infty{}^2 DC_L \cos(2\pi f_v t + \psi_0) \tag{7.33}$$

Here $f_v =$ vortex shedding frequency, which can be found from the Strouhal number, $St = f_v D/U_\infty$ and $C_L =$ amplitude of transverse force coefficient. ψ_0 is a constant, dependent on the history of the flow, that gives the phase of the transverse force. C_L and St are generally functions of Reynolds number.

For an oscillatory flow velocity

$$U_\infty = U_M \sin\left(\frac{2\pi t}{T}\right) \tag{7.34}$$

equation (7.33) is applied quasi-steady. The instantaneous vortex shedding frequency is proportional to the absolute value of the instantaneous velocity, i.e. to $|U_M \sin(2\pi t/T)|$. The magnitude of the instantaneous lift force is assumed proportional to the square of the instantaneous velocity. In finding the phase of the instantaneous lift force we have to generalize the expression $\cos(2\pi f_v t + \psi_0)$ for steady flow. This expression shows that during the time interval t to $t + dt$ there is an increase in the phase angle $d\psi = 2\pi f_v\, dt$. For unsteady flow this means

$$d\psi = 2\pi \frac{St}{D} \left| U_M \sin\left(\frac{2\pi t}{T}\right) \right| dt$$

By integration we can find the instantaneous value of ψ. For $0 \leqslant t/T < 0.5$ we can write

$$\psi - \psi_0 = \int_0^t d\psi = St \cdot KC\left(1 - \cos\left(\frac{2\pi t}{T}\right)\right)$$

During the second half cycle we can use the same expression if t is replaced by $t - \frac{1}{2}T$. We can now write the following quasi-steady model for the lift force.

$$F_L(t) = \frac{\rho}{2} U_M^2 \sin^2(2\pi t/T)\, C_L D$$

$$\times \cos\left[St \cdot KC\left(1 - \cos\left(\frac{2\pi t'}{T}\right)\right) + \psi_0 \right] \qquad (7.35)$$

where

$$t' = \begin{bmatrix} t & \text{for} & 0 \leqslant t/T < 0.5 \\ t - T/2 & \text{for} & 0.5 \leqslant t/T < 1.0 \end{bmatrix}$$

Further, C_L and St are lift coefficients and Strouhal number for steady flow. The equation is valid for any body shape where the steady lift force can be expressed by equation (7.33) and the ambient oscillatory flow velocity is given by equation (7.34). The phase angle ψ_0 in equation (7.35) need not be the same in different half cycles. A 180° phase change may occur.

EXPERIMENTAL TOOLS

Oscillatory flow tests in a U-tube

Sarpkaya has, in a series of publications (Sarpkaya & Isaacson, 1981), shown that the U-tube is an excellent experimental tool for measuring C_M, C_D, C_L and Strouhal number for plane harmonically oscillatory flow past two-dimensional cross-sections. Bearman and Graham have also used a U-tube extensively at Imperial College, London. A U-tube has also been built at the Norwegian Institute of Technology (Sortland, 1986). The Norwegian U-tube is described in Fig. 7.15. Forces in both lift and drag direction on cylinders of different shapes in oscillating flow can be measured. The force measurement system is illustrated in Fig. 7.16. Measurements on different configurations of cylinders like risers are also possible to achieve. The length of the measuring section can be changed in order to look at length-to-diameter ratio effects. The wall effect on the hydrodynamic coefficients can be simulated by placing the cylinders close to one of the walls. This is for instance important in studying sea floor effects on pipelines or slow-drift eddy-making coefficients for ship sections. In the latter case the free-surface effect can be approximated by a wall.

A drawback with the U-tube tests is scale effects, but this is the problem with most model testing of marine structures. Another drawback with U-tube testing is that it is difficult to simulate three-dimensional effects. On the other hand U-tube tests have provided and will provide a lot of physical insight in understanding vortex shedding in

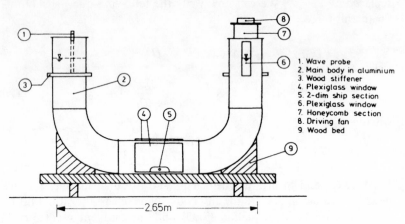

1. Wave probe
2. Main body in aluminium
3. Wood stiffener
4. Plexiglass window
5. 2-dim ship section
6. Plexiglass window
7. Honeycomb section
8. Driving fan
9. Wood bed

Fig. 7.15. Description of U-tube at Norwegian Institute of Technology. (Sortland, 1986.)

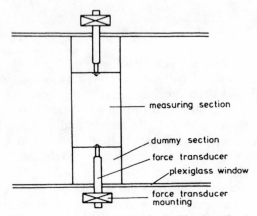

measuring section

dummy section
force transducer
plexiglass window

force transducer
mounting

Fig. 7.16. Force measurement system in the U-tube. (Sortland, 1986.)

oscillatory flow past bluff bodies. Flow visualization tests which are easy to do in the U-tube are also important for numerical calculations. Examples of useful flow visualization techniques are the 'hydrogen bubble', 'polystyrene particles' and Laser–Doppler techniques (Sortland, 1986).

The determination of C_M and C_D coefficients from U-tube tests is based on Morison's equation. For a circular cylinder this means that the in-line force F can be written as

$$F = \frac{\rho \pi D^2}{4} C_M L \dot{U} + \frac{\rho}{2} DLC_D U \, |U| \qquad (7.36)$$

where

D = diameter

L = length of test section

U = free-stream velocity

For oscillatory flow we can write

$$U = U_M \sin \frac{2\pi t}{T} \qquad (7.37)$$

According to Keulegan & Carpenter (1958) we can then write

$$C_D = \frac{3}{8} \frac{1}{\frac{1}{2}\rho U_M^2 LD} \int_0^{2\pi} F_M \sin \theta \, d\theta \qquad (7.38)$$

$$C_M = \frac{U_M T}{\pi^3 D} \frac{1}{\frac{1}{2}\rho U_M^2 LD} \int_0^{2\pi} F_M \cos \theta \, d\theta, \qquad \theta = \frac{2\pi t}{T} \qquad (7.39)$$

Here F_M is the measured force. In practice C_D and C_M will vary from cycle to cycle and in same cases one needs to average over 20 cycles to get good estimates of C_D and C_M. Fig. 7.17 shows an example of measured force signal and estimated force signal based on Morison's equation. Equations (7.38) and (7.39) have been used to determine C_D and C_M to give an optimum fit between the experiments and results of Morison's equation.

In a U-tube the period of oscillation T is constant. This corresponds to the natural period of the tank (see chapter 3 on U-tube anti-rolling tanks). The parameter we can vary is the velocity amplitude U_M. In a U-tube

$$\beta = \frac{\text{Rn}}{\text{KC}} = \frac{D^2}{vT}$$

is a constant for each model. One therefore often sees C_D and C_M presented as a function of KC for constant β. The results can easily be translated to be a function of both KC and Rn.

Free decay tests

It is common engineering practice to obtain damping coefficients from free decay tests. It is normal to assume that the motion can be written as

$$\ddot{x} + p_1 \dot{x} + p_2 |\dot{x}| \dot{x} + p_3 x = 0 \tag{7.40}$$

where p_1 is linear damping and p_2 is quadratic damping. Assuming the damping to be constant with respect to the amplitude of oscillation, the linear and quadratic damping coefficients p_1 and p_2 can be determined from the relation

$$\frac{2}{T_m} \log\left(\frac{X_{n-1}}{X_{n+1}}\right) = p_1 + \frac{16}{3}\frac{X_n}{T_m} p_2 \tag{7.41}$$

where X_n is the amplitude of the nth oscillation. There is one half period $T_m/2$ between X_n and X_{n+1} for any n. By plotting the left hand side of the equation versus $\frac{16}{3}X_n/T_m$ and fitting the points to a straight line by the least square method, the coefficients p_1 and p_2 are found; these give the linear and quadratic damping terms respectively, see Fig. 7.18.

It is difficult and in some cases impossible to determine this straight line from the experimental results. For instance, if the drag coefficient (i.e. the damping force) has a large KC-number dependence, or Reynolds-number dependence, it is impossible to find such a straight line. Problems may occur particularly if the oscillating system is not lightly damped. The reason is that the maximum velocity then changes significantly for each subsequent oscillation period. This means the Reynolds number and the KC-number change significantly. If the damping force is strongly dependent on these parameters, it means we

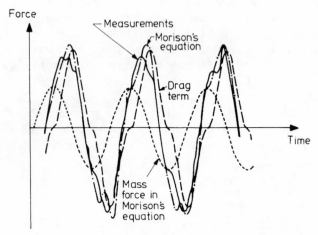

Fig. 7.17. Example of measured and estimated force by Morison's equation (equation (7.36)) from U-tube tests.

cannot find one p_1- and p_2-value that is valid for the total decay time. More than 10 oscillation periods with no significant variation in Reynolds number and KC-number may be needed to give a good estimate of the damping.

EXERCISES

7.1 Morison's equation

Consider deep-sea regular waves incident on a vertical rigid circular cylinder (Fig. 3.14). Consider the horizontal force on a strip of length dz (see equation (7.1)).

(a) Discuss at what time instants the maximum horizontal force F_{max} occurs as a function of the amplitude ζ_a and the wavelength λ of the incident waves. Derive expressions for F_{max} as a function of ζ_a and λ. Answer: With a_1 proportional to cos ωt:

   I) If $\zeta_a e^{kz} \leqslant 0.25\pi\, C_M D/C_D$ maximum force at $\omega t = 0$

   II) If $\zeta_a e^{kz} > 0.25\pi\, C_M D/C_D$ maximum force at cos $\omega t = 0.25\pi C_M D/(C_D \zeta_a e^{kz})$

(b) Discuss relative importance of the drag and inertia term as a function of KC-number.

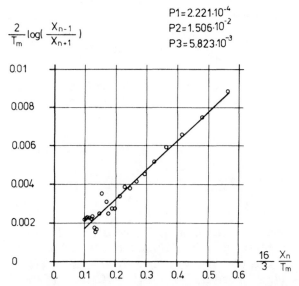

Fig. 7.18. Illustration of how the damping coefficients p_1 and p_2 (see equation (7.40)) are obtained from a free decay test. The free decay test is the calm water motion decay test in surge presented in Fig. 5.20. (The ordinate and the abscissa are defined in connection with equation (7.41).)

7.2 Lift-forces for high KC-number flow

(a) Equation (7.35) presents a formula for oscillatory lift forces on a two-dimensional body in planar oscillatory flow. Use the fact that $\cos(x \cos \phi)$ and $\sin(x \cos \phi)$ can be expressed by infinite series of Bessel's functions $\mathcal{J}_k$ of the first kind (Abramowitz & Stegun, 1964 (equations 9.1.44 and 9.1.45 on page 361)) to express the lift force in frequency components. Discuss how the frequency components depend on the KC-number. (Hint: It is possible to show that

$$F_{\rm L}(t) = {\rm Re}\bigg\{ 0.5\rho U_{\rm M}^2 C_{\rm L} D e^{i(z + \psi_0)}$$

$$\times \bigg[\mathcal{J}_0(z) + 2 \sum_{k=1}^{\infty} (-i)^k \mathcal{J}_k(z) \cos k\theta \bigg] \bigg\} \sin^2 \theta$$

where $z = KC \cdot St$, $\theta = 2\pi t/T$. This can be further rewritten in frequency components $k2\pi t/T$.)

(b) Study a fixed vertical circular cylinder in deep water regular waves. The cylinder is standing on the sea floor and penetrating the free-surface. Assume the vortex shedding is uncorrelated along the cylinder axis. Show how equation (7.35) can be applied to simulate the lift force distribution along the cylinder axis.

(c) Consider a vertical riser in deep water regular waves. Discuss 'lock-in' (vortex-induced oscillations).

7.3 Separated flow at small KC-numbers

Consider a fixed circular cylinder in planar ambient flow velocity.

$$U_x = U_{\rm M} \sin \omega t \tag{7.42}$$

(see Fig. 7.19). The velocity potential ϕ_0 for the non-separated flow can be written as

$$\phi_0 = U_{\rm M} \sin \omega t \left(r + \frac{R^2}{r} \right) \cos \theta \tag{7.43}$$

The KC-number is assumed small, but sufficiently large for the flow to separate. We will simulate the effect of the separated flow by introducing two time-dependent discrete vortices with circulations $\Gamma(t)$ and $-\Gamma(t)$ at the positions (r_v, θ_v) and $(r_v, -\theta_v)$. Fig. 7.19 shows that one has to add 'image' vortices inside the cylinder with circulations $-\Gamma$ and Γ at the positions $(R^2/r_v, \theta_v)$ and $(R^2/r_v, -\theta_v)$ in order to satisfy zero normal flow through the body surface. We will assume that a pair of vortices generated during one half cycle do not survive into the next half cycle. It

has therefore sufficient generality for us to study the flow in the time
interval $0 \leqslant \omega t \leqslant \pi$.

(a) Validate that the body boundary condition is satisfied when r_v
is close to R.

(b) Neglect the interaction effect of the two separated vortices on
each other and show that the angular velocity component of
the vortex at $\theta = \theta_v$ can be written as

$$V_0 = -\frac{\Gamma}{2\pi\left(r_v - \dfrac{R^2}{r_v}\right)} - U_M \sin \omega t \left(1 + \frac{R^2}{r_v^2}\right) \sin \theta \qquad (7.44)$$

(c) Calculate Γ by equation (6.14). Assume for simplicity when Γ
is integrated that the separation point (R, θ_s) is fixed and that
the influence of the separated vortices on the tangential
velocity U_{sa} in equation (6.14) can be neglected. (Answer:
$\Gamma = -U_M^2 \sin^2 \theta_s (\omega t - 0.5 \sin(2\omega t))/\omega$).

(d) Assume that a separated vortex follows the angular position of
the separation point. Assume the radial position r_v is close to
R and the motion of the separation point is negligible. Show
by means of equation (7.44) that this means

$$r_v - \frac{R^2}{r_v} \approx -\frac{\Gamma}{4\pi U_M \sin(\omega t) \sin \theta_s}$$

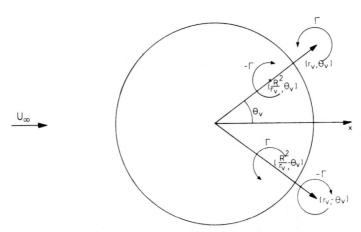

Fig. 7.19. Flow around a circular cylinder at small KC-numbers. The
separated flow is represented by two discrete vortices. ($\pm\Gamma =$
circulations of the shed vortices. Vortex positions (r_v, θ_v) and
$(r_v, -\theta_v)$.)

(e) From the unsteady Blasius equation (Graham, 1980) it is possible to show that the drag force F_1 due to the two separated vortices can be written as

$$F_1 = -\rho \sum_{n=1}^{2} \frac{\partial}{\partial t} (\Gamma_n (y - \hat{y}_n)) \tag{7.45}$$

where Γ_n is the circulation, and (x_n, y_n) is the position of vortex n. Further, $(\hat{x}_n, \hat{y}_n)$ is position of the image of vortex n inside the circle. Use previous results and evaluate the force at $\omega t = \pi/2$ (i.e. velocity maximum of incident flow). Show that

$$C_D = \left(\frac{KC}{\pi} \right) \sin^4 \theta_s \tag{7.46}$$

Assume reasonable values of θ_s and compare the results with equation (7.24).

8 STATIONKEEPING

Precise position and motion control of ships and other floating structures is important from a marine operational point of view. Thrusters and mooring systems are important means of holding a structure against wind, waves and current. In earlier chapters we have discussed how to evaluate environmental forces due to waves and current. Wind has not been covered as extensively as current. One of the reasons is that there are many similarities between wind and current loads. The most important differences are perhaps in the differences in magnitude of density and fluid velocity in the two media. From a fluid mechanics point of view one may also say that the structural parts above sea level are generally more complicated than the submerged parts. It is in any case necessary to rely on model tests to obtain current and wind loads on marine structures. One has more confidence in using calculations for mean wave loads. This is particularly true for large-volume structures if one uses an accurate method based on, for instance, three-dimensional source technique or Green's second identity to evaluate the first-order fluid flow.

In this chapter we will discuss forces from mooring systems and thruster systems.

MOORING SYSTEMS

A mooring system is made of a number of cables which are attached to the floating structure at different points with the lower ends of the cables anchored at the sea bed. One type is the vertical tension leg mooring which is used in connection with tension leg platforms. This is a moored stable platform for which the buoyancy exceeds the platform weight, and the net equilibrating vertical force is supplied by vertical tension mooring cables secured by deadweight or drilled-in anchors. The mooring lines provide essentially total restraint against vertical movement of their upper ends.

In a spread mooring system, several pre-tensioned anchor lines are arrayed around the structure to hold it in the desired location. The normal case is that the anchors can be easily moved. This implies that the anchor in operation cannot be loaded by too large vertical forces, and, to ensure that the anchors are kept in position, it is necessary that a significant part of the anchor lines lie on the seabed. The cables are made up of either chain, rope or a combination of both. The ropes are available in constructions from steel, natural fibre and synthetic fibres. Segmented anchor lines, i.e. cables composed of two or more lengths of different material, are used to get a heavy cable at the bottom (i.e. chain), and a light line close to the water surface. This gives greater stiffness and lighter anchor lines, than the use of chain or wire alone. The tension forces in the cables, which are the means of applying restraining forces on the floating structure, are due to the cable weight and/or its elastic properties, depending on the manner in which the cable system is laid.

The initial tension, or pre-tension in a cable is often established by the use of winches on the vessel or platform (often called a floating unit in offshore connection). The winches pull on the cables to establish the desired cable configuration. As the unit moves in response to unsteady environmental loads, the tension in the cables changes due to varying cable geometry. Thus the mooring cables have an effective stiffness composed of an elastic and a geometric stiffness, which, combined with the motions of the unit, introduce forces depending on the mooring cable characteristics.

In the following text we will show how to analyse a cable line from a static point of view.

Static analysis of a cable line

A picture of the anchor line is shown in Fig. 8.1. We assume a horizontal sea bed. The cable is in a vertical plane coinciding with x–z-plane. We neglect bending stiffness which is a good approximation for chains. It is also appropriate for wires with a large radius of curvature. We neglect dynamic effects in the line.

Fig. 8.2 shows one element of the cable line. Forces D and F acting on the element are the mean hydrodynamic forces per unit length in the normal and tangential direction respectively. w is the weight per unit length of the line in water, A is the cross-sectional area of the cable line, E is the elastic modulus and T is the line tension. Because w is the weight in water, this introduces correction forces, $-\rho gAz$ and $-\rho gAz - \rho gA\,dz$ at the ends of the element. In this way we can calculate correctly the hydrostatic forces on the element.

From Fig. 8.2 we find that

$$dT - \rho gA\,dz = [w \sin \phi - F(1 + T/(AE))]\,ds$$

$$T\,d\phi - \rho gAz\,d\phi = [w \cos \phi + D(1 + T/(AE))]\,ds$$

These equations are non-linear and it is in general not possible to find an explicit solution. However, for many operations it is a good approximation to neglect the effect of the current forces F and D. We will also

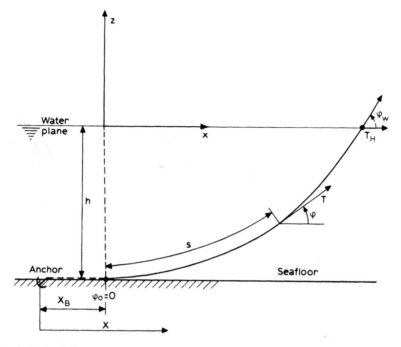

Fig. 8.1. Cable line with symbols.

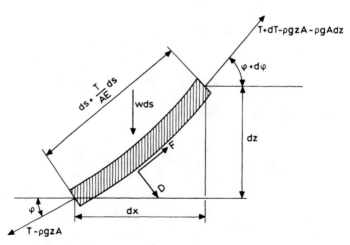

Fig. 8.2. Forces acting on an element of an anchor line.

neglect the effect of elasticity. This simplifies the analysis. However, in extreme conditions elasticity should be accounted for. We will later show how this can be done in an approximate way. We will assume the cable line has constant weight per unit length.

Solutions of the inelastic cable line (catenary) equations
By introducing

$$T' = T - \rho g z A \tag{8.1}$$

we can write

$$dT' = w \sin \phi \, ds \tag{8.2}$$

$$T' \, d\phi = w \cos \phi \, ds \tag{8.3}$$

By dividing these two equations we see that

$$\frac{dT'}{T'} = \frac{\sin \phi}{\cos \phi} d\phi$$

i.e.

$$T' = T_0' \frac{\cos \phi_0}{\cos \phi} \tag{8.4}$$

By integrating equation (8.3) we find that

$$s - s_0 = \frac{1}{w} \int_{\phi_0}^{\phi} \frac{T_0'}{\cos \theta} \frac{\cos \phi_0}{\cos \theta} d\theta = \frac{T_0' \cos \phi_0}{w} [\tan \phi - \tan \phi_0] \tag{8.5}$$

Since $dx = \cos \phi \, ds$ we can write

$$\begin{aligned}
x - x_0 &= \frac{1}{w} \int_{\phi_0}^{\phi} \frac{T_0' \cos \phi_0}{\cos \theta} d\theta \\
&= \frac{T_0' \cos \phi_0}{w} \left(\log\left(\frac{1}{\cos \phi} + \tan \phi \right) \right. \\
&\quad \left. - \log\left(\frac{1}{\cos \phi_0} + \tan \phi_0 \right) \right)
\end{aligned} \tag{8.6}$$

Since $dz = \sin \phi \, ds$ we find that

$$\begin{aligned}
z - z_0 &= \frac{1}{w} \int_{\phi_0}^{\phi} \frac{T_0' \cos \phi_0 \sin \theta}{\cos^2 \theta} d\theta \\
&= \frac{T_0' \cos \phi_0}{w} \left[\frac{1}{\cos \phi} - \frac{1}{\cos \phi_0} \right]
\end{aligned} \tag{8.7}$$

We choose ϕ_0 to be the point of contact between the cable line and the sea bed, i.e. $\phi_0 = 0$. From equation (8.4) we see that

$$T_0' = T' \cos \phi \tag{8.8}$$

The horizontal component of the tension at the waterplane may be written as

$$T_H = T \cos \phi_w \tag{8.9}$$

By comparing equations (8.1), (8.8) and (8.9) we find that

$$T_0' = T_H$$

This is also consistent with a global balance of forces. By the way, we have chosen the coordinate system $x_0 = 0$ and $z_0 = -h$. Further, we set $s_0 = 0$. The angle ϕ can be eliminated from equation (8.5) and (8.7) by equation (8.6) which can be written as

$$\frac{xw}{T_H} = \log\left(\frac{1 + \sin \phi}{\cos \phi}\right)$$

i.e.

$$\sinh\left(\frac{wx}{T_H}\right) = \frac{1}{2}\left(\frac{1 + \sin \phi}{\cos \phi} - \frac{\cos \phi}{1 + \sin \phi}\right) = \tan \phi$$

$$\cosh\left(\frac{wx}{T_H}\right) = \frac{1}{2}\left(\frac{1 + \sin \phi}{\cos \phi} + \frac{\cos \phi}{1 + \sin \phi}\right) = \frac{1}{\cos \phi}$$

We may now write

$$s = \frac{T_H}{w} \sinh\left(\frac{w}{T_H} x\right) \tag{8.10}$$

$$z + h = \frac{T_H}{w}\left[\cosh\left(\frac{w}{T_H} x\right) - 1\right] \tag{8.11}$$

The line tension can be found by combining equations (8.1), (8.7) and (8.8), i.e.

$$T - \rho g z A = \frac{T_H}{\cos \phi} = T_H + w(z + h)$$

i.e.

$$T = T_H + wh + (w + \rho g A)z \tag{8.12}$$

The vertical component T_z of the tension is found by using

$$dT_z' = d(T' \sin \phi) = dT' \sin \phi + T' \cos \phi \, d\phi$$
$$= w \sin^2 \phi \, ds + w \cos^2 \phi \, ds$$

This means $T_z' = ws$, which in the waterplane says

$$T_z = ws \tag{8.13}$$

We will show how we can use the formulas in a practical context. Let us consider the example illustrated in Fig. 8.3 and try to find the minimum length l_{min} of the chain. We will assume gravity anchors are used. These are commonly used when floating units like ships and drilling platforms are moored. A requirement is that a gravity anchor cannot be exposed to vertical forces from the anchor lines. We will use this to determine l_{min}. However, it should be noted that for practical design cases the line length is determined from a set of different conditions considering both an intact mooring system and broken line situations.

To find the minimum length of the cable lines we will first use equations (8.10) and (8.11), i.e.

$$l_s = a \sinh\left(\frac{x}{a}\right) \tag{8.14}$$

$$h = a\left[\cosh\left(\frac{x}{a}\right) - 1\right] \tag{8.15}$$

where

$$a = \frac{T_H}{w} \tag{8.16}$$

By combining equations (8.14) and (8.15) we see that

$$l_s^2 = h^2 + 2ha \tag{8.17}$$

From equation (8.12) we see that the maximum tension in the cable line can be written as

$$T_{max} = T_H + wh \tag{8.18}$$

By combining equations (8.15), (8.17) and (8.18) we see that the minimum length of the cable is

$$l_{min} = h\left(2\frac{T_{max}}{wh} - 1\right)^{\frac{1}{2}} \tag{8.19}$$

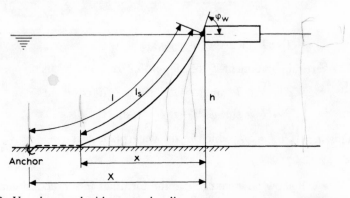

Fig. 8.3. Vessel moored with one anchor line.

We will set T_{max} equal to T_{br}, i.e. the breaking strength of the cable. For instance if $T_{br} = 1510\,kN$, $w = 828\,N/m$ and $h = 25\,m$ we find that $l_{min} = 301\,m$.

If we want to find the mean position of the vessel in wind, waves and current we have to know the horizontal force T_H from the cable on the vessel as a function of the horizontal distance X between the anchor and the point where the anchor line is connected to the vessel. We can write the horizontal distance X as (see Fig. 8.3)

$$X = l - l_s + x \qquad (8.20)$$

By using equation (8.17) to express l_s and equation (8.15) to express x we see the following relation between X and T_H

$$X = l - h\left(1 + 2\frac{a}{h}\right)^{\frac{1}{2}} + a\,\cosh^{-1}\left(1 + \frac{h}{a}\right) \qquad (8.21)$$

where $a = T_H/w$.

Calculations based on equation (8.21) are illustrated in Fig. 8.4. The weight per unit length of the chain in water is $w = 828\,N\,m^{-1}$, the water depth $h = 25\,m$ and the length l of the chain outside the ship is 100 m.

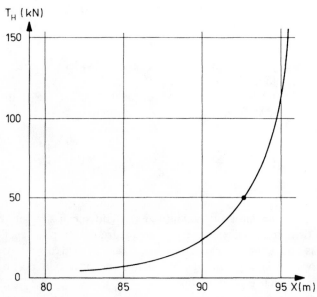

Fig. 8.4. Example of the horizontal force from an anchor line on a vessel as a function of the horizontal distance X between the anchor and the point where the anchor line is connected to the vessel. (The vessel and the anchor line configuration is shown in Fig. 8.3.) Water depth: 25 m. Weight per unit length of chain in water: $828\,N\,m^{-1}$. Chain length: 100 m.

To find X for a vessel that is moored with a mooring system as described above we have to know the environmental forces on the vessel. Let us say that the x-component of the *average* wave, current and wind force on the vessel was 50 kN. From Figure 8.4 or equation (8.21) we find that the average distance X from the anchor line to the point where the anchor line is connected to the vessel will be 93 m. However, due to the vessel's motions in waves the distance X will oscillate around 93 m. From the figure we note that the horizontal force from the anchor line on the vessel will also oscillate. If the horizontal motions are not too large, we may write

$$T_H = (T_H)_M + C_{11}\eta_1$$

Here $(T_H)_M$ is the average horizontal force from the anchor line on the vessel, i.e. 50 kN in our example, and η_1 is the horizontal motion in the x-direction of the point on the vessel where the cable line is connected to the vessel. We can find C_{11} from Fig. 8.4 as the derivative of T_H with respect to X at $(T_H)_M$. We may also find an analytical expression of C_{11}. By differentiating equation (8.21) it follows that

$$C_{11} = \frac{dT_H}{dX} = w\left[\frac{-2}{\left(1 + 2\frac{a}{h}\right)^{\frac{1}{2}}} + \cosh^{-1}\left(1 + \frac{h}{a}\right)\right]^{-1} \tag{8.22}$$

where $a = (T_H)_M/w$. From this example it is evident that the anchor line has a spring effect on the vessel.

If we denote the horizontal motion in the x-direction of the centre of gravity of the ship as η_1 we may approximate the surge equation of motion of the ship as

$$(M + A_{11})\frac{d^2\eta_1}{dt^2} + B_{11}\frac{d\eta_1}{dt} + B_D\left|\frac{d\eta_1}{dt}\right|\frac{d\eta_1}{dt} + C_{11}\eta_1 = F_1(t) \tag{8.23}$$

where M is the mass of the ship, A_{11} is added mass in surge, B_{11} is a linear damping coefficient, B_D is a quadratic damping coefficient and $F_1(t)$ is the dynamic excitation force due to waves and wind. From equation (8.23) we find that resonance in surge will occur when the circular frequency of oscillation

$$\omega = \left(\frac{C_{11}}{M + A_{11}}\right)^{\frac{1}{2}} \tag{8.24}$$

In a practical case resonance oscillations in surge may be excited by slowly-varying excitation forces due to waves and wind. The most important linear damping B_{11} for slow-drift oscillation of a large volume structure is due to the wave-drift damping (see chapter 5). The quadratic damping is due to drag forces on the hull (see chapter 7) and drag damping from the anchor lines. Strictly speaking we should have included the effect of the relative velocity between the vessel and linear incident wave field when we formulated equation (8.23).

Analysis of a spread mooring system

The above procedure for one cable line can be generalized to a spread mooring system consisting of several cable lines. The relationship between mean external loads on the vessel and its position can be found by considering contributions from each cable line separately. We can write the horizontal forces F_1^M, F_2^M and the yaw moment F_6^M from the mooring lines as

$$F_1^M = \sum_{i=1}^{n} T_{Hi} \cos \psi_i$$

$$F_2^M = \sum_{i=1}^{n} T_{Hi} \sin \psi_i \qquad (8.25)$$

$$F_6^M = \sum_{i=1}^{n} T_{Hi}[x_i \sin \psi_i - y_i \cos \psi_i]$$

Here T_{Hi} is the horizontal force from anchor line number i. Its direction is from the attachment point of the anchor line towards the anchor. Further, x_i and y_i are respectively the x- and y-coordinate of the attachment point of the anchor line to the vessel and ψ_i is the angle between the anchor line and the x-axis as defined in Fig. 8.5. In order

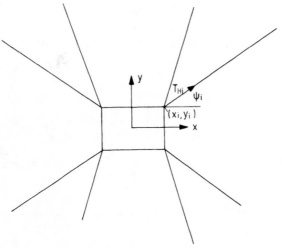

Fig. 8.5. Example of spread mooring system for a drilling platform.

for the moored structure to be in equilibrium, F_1^M, F_2^M and F_6^M have to balance the mean forces due to waves, wind and current. The problem of finding the equilibrium position can generally not be solved directly and an iteration procedure has to be set up.

If we want to find the linear restoring effect of the anchor lines in the equations of motion we can generalize the procedure outlined in connection with equation (8.22). We can write

$$C_{11} = \sum_{i=1}^{n} k_i \cos^2 \psi_i$$

$$C_{22} = \sum_{i=1}^{n} k_i \sin^2 \psi_i$$

$$\text{(8.26)}$$

$$C_{66} = \sum_{i=1}^{n} k_i (x_i \sin \psi_i - y_i \cos \psi_i)^2$$

$$C_{26} = C_{62} = \sum_{i=1}^{n} k_i (x_i \sin \psi_i - y_i \cos \psi_i) \sin \psi_i$$

Here k_i is the restoring coefficient for anchor line number i. It can be found as in equation (8.22). The coupling coefficients C_{16}, C_{61}, C_{12} and C_{21} are zero if the mooring arrangement is symmetric about the x–z plane.

Example. Static analysis of a spread mooring system

A platform is restrained from moving by two anchors with anchor lines of chain (see Fig. 8.6). The length l of the anchor line is 600 m and the horizontal pre-tension T_H is 300 kN. The weight of the chain in water is $w = 1000 \text{ N m}^{-1}$. Assume static environmental forces on the platform.

(a) What is the distance between anchors A and B?

(b) What is the line tension at the platform?

(c) Find out how large the environmental force on the platform has to be to move anchor A.

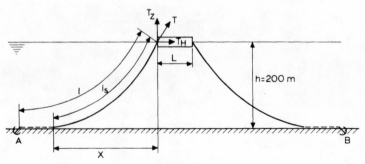

Fig. 8.6. Vessel moored with two anchor lines.

(d) Study the displacement of the platform as a function of the environmental force on the platform.

Solution:

(a) The distance between anchors A and B is given as $2X + L$ where $X = l - l_s + x$ and L is the length of the platform. By using equation (8.21) we find that the distance between the two anchors is

$$1060 \text{ (m)} + L$$

(b) From equation (8.12) it follows that the line tension at the platform is

$$T = T_H + wh = 500 \text{ kN}$$

(c) and (d) Let us first examine mooring line A. Anchor A will start moving if $l_s = l = 600$ m. From equations (8.16) and (8.17) we will find the corresponding horizontal tension T_H in cable line A and the x-value for attachment point of the cable line to the platform (see Fig. 8.3).
We find $a = 800$ m, $T_H = 800$ kN and

$$x = 800 \sinh^{-1}\left(\frac{600}{800}\right) = 555 \text{ m}$$

This means the attachment point of the cable line to the platform has moved 25 m compared to the original situation. In order to find how large the environmental force on the platform has to be to move the platform 25 m, we will first study the horizontal tension T_H in each cable line as a function of displacement. This is illustrated in Fig. 8.7 and can be performed in the same way as described for the vessel shown in Fig. 8.3. Instead of referring the displacement to one of the anchor positions, we use the displacement of the vessel relative to the position that the centre of gravity of the vessel had when there were no environmental forces acting on the vessel, i.e. the situation described in the introduction of the example. From Fig. 8.7 we find that the horizontal tension in cable line A is 800 kN and the horizontal tension in cable line B is 140 kN when the vessel has moved 25 m to left. From this we may conclude that the necessary horizontal environmental force on the platform in direction A–B to move anchor A is $800 - 140 = 660$ kN. We may also use Fig. 8.7 to study the displacement of the platform as a function of the environmental forces on the platform. This may be done by giving the platform a displacement δ and finding the horizontal tension in line A (T_{HA}) and horizontal tension in line B (T_{HB}) from Fig. 8.7. The horizontal environmental force in direction A–B has to be $T_{HA} - T_{HB}$ to get a balance of the forces acting on the platform. The functional relationship between $T_{HA} - T_{HB}$ and δ is plotted in Fig. 8.8.

Solution of the elastic cable line equations

In extreme conditions we have to consider the effect of the elasticity in the cable lines. We will show how this can be done in an approximate way. Equations (8.2) and (8.3) which set up relationships between the effective tension T, element angle ϕ and unstretched element length ds are still valid. This means equations (8.4) and (8.5) are also valid. Since we have chosen ϕ_0 and s_0 to be zero, it means that

$$T' = \frac{T_0'}{\cos \phi} \tag{8.27}$$

$$s = \frac{T_0'}{w} \tan \phi \tag{8.28}$$

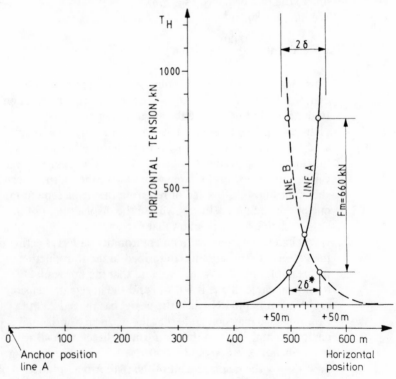

Fig. 8.7. Horizontal force on the vessel from anchor lines A and B as a function of horizontal displacement of the vessel illustrated in Fig. 8.6.

To find the x- and z-coordinates we have to use the following relationship between the stretched length dp and unstretched length ds of a cable element

$$dp = ds\left(1 + \frac{T}{AE}\right) \qquad (8.29)$$

This means

$$\frac{dx}{ds} = \cos\phi\left(1 + \frac{T}{AE}\right) \approx \cos\phi\left(1 + \frac{T'}{AE}\right) = \cos\phi + \frac{T_0'}{AE} \qquad (8.30)$$

$$\frac{dz}{ds} = \sin\phi\left(1 + \frac{T}{AE}\right) \approx \sin\phi\left(1 + \frac{T'}{AE}\right) = \sin\phi + \frac{w}{AE}s \qquad (8.31)$$

The last term on the right hand side of equation (8.31) follows by combining equation (8.27) with equation (8.28). We will now integrate equations (8.30) and (8.31) up to the attachment point of the cable line to the vessel. For simplicity we will assume the attachment point is at $z = 0$. We should note that the vertical component T_z of the tension at the attachment point is given by equation (8.13). This means the unstretched length l_s (see Fig. 8.3) of the cable from the point where cable first touches the bottom to the attachment point can be written

$$l_s = \frac{T_z}{w} \qquad (8.32)$$

By integrating equation (8.31) in a similar way as in equation (8.7) it follows that

$$h = \frac{T_H}{w}\left[\frac{1}{\cos\phi_w} - 1\right] + \frac{1}{2}\frac{w}{AE}l_s^2$$

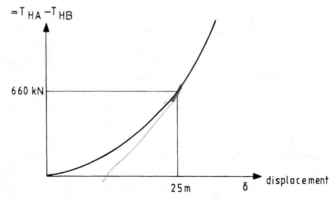

Fig. 8.8. Mean environmental horizontal force on the vessel described in Fig. 8.6 as a function of horizontal displacement.

where $\cos \phi_w$ at the attachment point can be written $T_H/(T_H^2 + T_z^2)^{\frac{1}{2}}$. From this equation we can find an explicit expression for T_H, i.e.

$$T_H = \frac{T_z^2 - \left(wh - \dfrac{1}{2}\dfrac{w^2}{AE}l_s^2\right)^2}{2\left(wh - \dfrac{1}{2}\dfrac{w^2}{AE}l_s^2\right)} \tag{8.33}$$

We should note that the tension T at the attachment point is simply

$$T = (T_H^2 + T_z^2)^{\frac{1}{2}} \tag{8.34}$$

Finally, by integrating equation (8.30) in a similar way to equation (8.6) we find that

$$x = \frac{T_H}{w}\log\left(\frac{(T_H^2 + T_z^2)^{\frac{1}{2}} + T_z}{T_H}\right) + \frac{T_H}{AE}l_s \tag{8.35}$$

We can now use equations (8.32) to (8.35) in the following way. We start out by assuming a value of T_z, then calculate l_s from equation (8.32), T_H from equation (8.33), T from equation (8.34) and finally x from equation (8.35).

If we want to find the mooring line characteristics for a given value of T_H, we have to perform the calculation procedure above for several assumed values of T_z and then afterwards interpolate the data.

The equations that we have found for cable lines can also be used for analysis of towing cables. The water depth in the formulas have then to be replaced by the sag s_d of the cable. Further, l_s is the same as half the cable length.

THRUSTER FORCES

Thrusters may be used in combination with a mooring system or alone to keep a vessel in position. From open water testing one can get a first estimate of the thruster performance, but a thruster may lose efficiency due to interaction with other thrusters, the hull, current and waves. One example of the hull interaction effect is the Coanda effect. This means that the propeller (thruster) slip stream is attracted by the hull (see Fig. 8.9). A consequence of this is loss of thrust. The Coanda effect can be explained by representing the propeller slip stream by a circular jet, which acts like a line of sinks. This means that the water is entrained in the jet from outside the jet. (For analysis of circular jet flow see Schlichting, 1979: pp. 747–50). If the boundary had not been there, the entrainment velocity would have been directed radially toward the centre of the jet. The entrained velocity is only a function of the radius. If a boundary is present, the velocity will be at a maximum between the jet and the boundary. High velocity means low pressure. This means a

pressure difference across the jet with a resultant force towards the
boundary. The force will attract the jet towards the wall. The attraction
force has to be in balance with the centrifugal force of the jet flow. This
determines the position of the jet relative to the wall. This is illustrated in
Fig. 8.10. For a thin jet initially at a distance h from an infinitely long
wall, it takes roughly $6 \cdot h$ to hit the wall if the jet is initially parallel to
the plane wall. This information can be used as a rough tool for avoiding
the propeller slip stream coming into contact with the hull. However, we
should realize that it is an idealization to approximate the propeller slip
stream by a thin jet. The jet must have a radius comparable to the
propeller radius. Close to the propeller the flow is not jet-like. It actually
takes a distance of about six times the diameter from the thruster before
the propeller slip stream develops into a fully turbulent jet-like flow. The
jet will spread as a function of the distance from the thruster. If we
consider a free jet, i.e. a jet not in presence of a boundary, the points in a
jet where the velocity is half the maximum velocity will spread with an
angle of about 5°. This spreading may cause the propeller slip stream to
be in contact with the hull before the above attraction effect has fully
developed. When the propeller stream clings to the hull, it behaves like a
wall jet. The wall jet may very well separate from the hull again. This
depends on the local radius of curvature of the hull. If a sharp corner is
present the propeller stream will separate and there will be no significant
thrust loss. It is difficult to estimate theoretically what the loss due to the

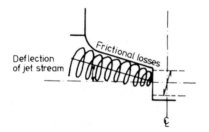

Fig. 8.9. Stern skeg tunnel thruster. Example of Coanda effect.

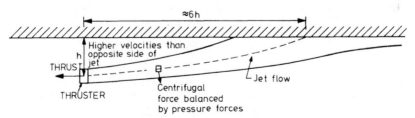

Fig. 8.10. Deflection of propeller slip stream. (Coanda effect.)

Coanda effect is. Limited full scale experience from a supply ship indicates that the Coanda effect may cause a 30–40% loss of thrust for a given power. In that case the propeller stream followed the ship hull all the way up to the free-surface. If the propeller stream had separated from some point on the hull surface, it is expected that this would have caused a smaller loss. For semi-submersibles the Coanda effect for a thruster on one pontoon may cause the propeller stream to hit another pontoon. The loss due to this can roughly be estimated by considering the propeller stream to be an incident current on the other pontoon. Another case for a semi-submersible would be if the thruster is aligned longitudinally along a pontoon. Due to the long distance the propeller slip stream is likely to be completely attracted to the pontoon. The boundary flow between the wall jet and the pontoon will cause shear forces. This may amount to 10–15% loss of power.

Loss of efficiency due to current
The forces generated by a tunnel thruster are affected by the flow of a current past the entrance and exit of the tunnel. Model test results taken from Chislett & Bjørheden (1966) and presented in Fig. 8.11 show the percentage loss of side force as a function of vessel forward speed.

The loss of thrust is caused by deflection of the jet stream and by

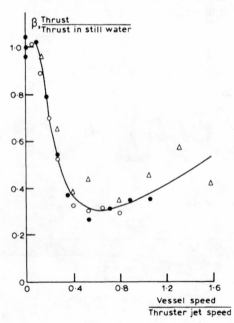

Fig. 8.11. Bow thruster performance. Effect of ahead current/vessel speed. (Chislett & Bjørheden, 1966.)

interactions between the jet stream and the hull which give rise to suction forces on the tunnel exit side of the vessel.

In order to illustrate the results in Fig. 8.11, we may evaluate the propeller jet velocity V_j by

$$V_j = \left(\frac{T}{\rho A_0}\right)^{\frac{1}{2}} \tag{8.36}$$

where T is the thrust and A_0 is the propeller disk area. This follows from conservation of momentum in the fluid. For instance, with $T = 130\,\text{kN}$, $A_0 = 3.5\,\text{m}^2$, $V_j = 6\,\text{m s}^{-1}$. A current speed of $1\,\text{m s}^{-1}$ along the ship centerline means that the thrust of a thruster working in the transverse direction of the ship is $\approx 80\%$ of the still water thrust.

Minsaas et al. (1986) have argued that a similar effect to this must be present in waves. They did experiments with a fictitious bow thruster system in head waves. The ship sides were simulated by vertical plates parallel to the incident regular waves. The whole system was restrained from oscillating. At the propeller centre the wave velocity amplitude was written as

$$V_w = \omega \zeta_a e^{-k h_0} \tag{8.37}$$

where h_0 is the still water submergence of the propeller shaft.

By interpreting V_w as a current velocity they were able to predict trends in their experimental results. If a ship is moving in waves the problem becomes more complicated. The velocity across the propeller jet at the tunnel entrance is no longer V_w. This problem area requires further research.

Influence of free-surface effects on thruster characteristics

In this section we will talk about other effects of the wave-induced motion on the thruster behaviour. These are associated with the thruster coming close to the free-surface. Depending on the thruster loading this may cause air ventilation with a serious loss of propeller thrust and torque.

In order to study the effects of waves, tests in calm water were performed by Minsaas et al. (1986), based on a quasi-steady assumption. As long as the thruster is not ventilating, this is legitimate, since the wave-induced motions occur with a much lower frequency than the propeller rotation. However, when the propeller is in a transient ventilating condition, there is also a frequency connected with the development of the ventilated area on the propeller blades. This wave-induced effect cannot be simulated by means of results in calm water.

The tests in calm water were performed with different propeller axis submergences h relative to the free-surface. By the quasi-steady assumption one can interpret h as the submergence of the instantaneous position of the propeller axis below the instantaneous position of the wave surface. (see Fig. 8.12). By averaging the propeller thrust and moments in time one can find the effect of the wave-induced motions of the ship on the thruster characteristics.

Typical model test results from calm water are shown in Fig. 8.13 where the thrust is presented as a fraction of the thrust for a fully immersed propeller:

$$\beta_T = \frac{K_T(h/R)}{K_{T_0}} \tag{8.38}$$

and for different immersions and number of revolutions n. R means the

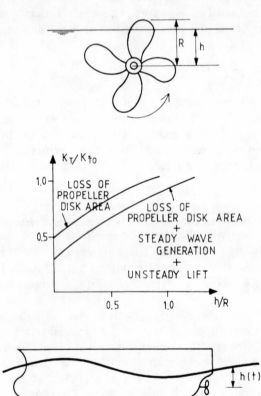

Fig. 8.12. Quasi-steady approximation of the effect of waves on propeller characteristics when ventilation does not occur. (K_T/K_{T_0} = the thrust as a fraction of the thrust for a fully immersed propeller.)

propeller radius. It is observed that the number of revolutions has a marked influence on the results. The effect of loss of effective propeller area is important. When $h/R < 1.0$ the ratio of the immersed disk area A_1 and the total disk area of the propeller is:

$$\beta_0 = \frac{A_1}{A_0} \qquad (8.39)$$

If we assume that the thrust is proportional to A_1, β_0 gives the thrust reduction for small n-values as illustrated in Fig. 8.13. For large n-values we see that the thrust is very low for the small h/R-values. This is due to propeller ventilation. Qualitatively it can be explained as follows. Increased n means increased loading. This means large suction pressures on the propeller. The lower the pressure is, the more likely it is that ventilation occurs. For $h/R = 1.0–1.5$ there is a very rapid variation in thrust. This is when ventilation starts. The behaviour of β in waves is not quasi-steady when the propeller is in a transient ventilating state. The maximum thrust is reached later in time than maximum immersion. In most of the cases maximum thrust never reaches the same thrust value as when the propeller was fully immersed in static conditions.

Minsaas et al. (1986) have tried to apply their experimental results for a ducted propeller and bow thruster in regular waves to thrust loss for different ships and sea states. An example for a 250 m long ship is presented in Fig. 8.14. The results will depend on, for instance, propeller shaft submergence, propeller diameter, propeller pitch, propeller revolution and hull form. The results in Fig. 8.14 show that even a large ship experiences considerable thrust losses in rough sea.

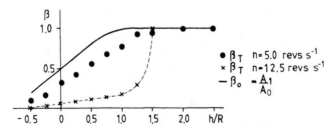

Fig. 8.13. Loss of thrust due to reduced immersion of a bow thruster in calm water. Model scale. (β_T = thrust divided by thrust for deeply submerged propeller. β_0 (see equation (8.39)), n = number of propeller revolutions per second, h = propeller axis submergence relative to the free surface, R = propeller radius.) (Minsaas et al., 1986.)

THRUSTER PERFORMANCE AND DYNAMIC POSITIONING

If the thrusters are part of a dynamic positioning (DP) system, an idealized simplification of the total thruster forces on a structure can be written as

$$F_k = \bar{F}_k - \sum_{j=1}^{6} (B_{kj}{}^{\text{DP}}\dot{\eta}_j + C_{kj}{}^{\text{DP}}\eta_j) \qquad k = 1, \ldots, 6 \qquad (8.40)$$

Here $k = 1$ means surge force, $k = 2$ sway force, $k = 3$ heave force, $k = 4$ roll moment, $k = 5$ pitch moment and $k = 6$ yaw moment. For a dynamically positioned ship it is only $k = 1, 2$ and 6 that are of interest. $\bar{F}_k$ means mean forces. They have to balance the mean wave, current and wind loads. Further, η_j are the slowly-varying motions of the structure, obtained through proper filtering of the motion reference measurements. It is the high-frequency motion due to waves that are filtered out. It is generally impossible to have a system that can react to the high-frequency wave forces. To get a feeling for the values of the damping matrix $B_{kj}{}^{\text{DP}}$ and the restoring matrix $C_{kj}{}^{\text{DP}}$, we can refer to a dynamically positioned ship. As a first guess we can set the coupling coefficients equal to zero. Further, we can choose $C_{kk}{}^{\text{DP}}$, $k = 1, 2$ and 6 so that the natural period of the slowly oscillating ship is from 100–200 s in surge, sway and yaw. The damping coefficients $B_{kk}{}^{\text{DP}}$ can be set equal to $\approx 60\%$ of the critical damping, for motion mode k. The final real values are decided after sea tests of the DP system.

Let us show what the first guess implies for the surge motion. We can

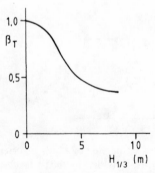

Fig. 8.14. An example of thrust reduction for a bow thruster on a 250 m long ship in head sea waves. No current. (β_{T} = thrust divided by thrust for deeply submerged propeller, $H_{\frac{1}{3}}$ = significant wave height.) (Minsaas *et al.*, 1986.)

write the equation of slow-drift surge motion as

$$(M + A_{11})\frac{d^2\eta_1}{dt^2} + (B_{11} + B_{11}{}^{DP})\frac{d\eta_1}{dt} + C_{11}{}^{DP}\eta_1$$

$$= F_1{}^{SW} + F_1{}^{wind} \quad (8.41)$$

If the ship has a mooring system, its effect on equation (8.41) can be taken care of by adding a restoring term $C_{11}\eta_1$. In equation (8.41) M means the ship mass, A_{11} the surge added mass, B_{11} the hydrodynamic damping, $F_1{}^{SW}$ the slow-drift wave excitation force and $F_1{}^{wind}$ gust excitation force. The mean values of $F_1{}^{SW}$ and $F_1{}^{wind}$ are zero. According to what we said just recently we should choose

$$C_{11}{}^{DP} = \left(\frac{2\pi}{T_n}\right)^2 (M + A_{11}) \quad \text{where } T_n = 100 \text{ s for instance}$$

$$B_{11}{}^{DP} = 1.2[(M + A_{11})C_{11}{}^{DP}]^{\frac{1}{2}} \quad (8.42)$$

The variance $\sigma_{\eta 1}$ of the slow-drift surge motion can be calculated as we did for the slow-drift motion in chapter 5 (see equation (5.47)). We can also calculate the variance of the total thruster surge force by using equation (8.40).

We should note that we have all the time talked about the total thruster forces. For a DP system there has to be an allocation system that tells how the power should be distributed among the individual thruster units.

EXERCISES

8.1 Static analysis of a cable line
Given:

Weight of chain in water	$w = 828 \text{ N m}^{-1}$
Breaking strength of the chain	$T_{br} = 1510 \text{ kN}$
Water depth	$h = 25 \text{ m}$
Length of chain outside the ship	$l = 100 \text{ m}$

Further, the cable line is connected to the vessel just above the waterplane. Assume in questions (a), (b) and (c) that the cable line is inelastic.

    (a) Find out the necessary horizontal force T_H from the vessel on the chain in order to be able to move the anchor. (Answer; 155 250 N).

    (b) What is the breaking safety of the chain in the condition described in question (a)? (Answer: 8.6.)

(c) Calculate the angle θ_w between the cable line and the mean water surface as a function of the horizontal distance X between the anchor and the point where the anchor line is connected to the vessel (see Fig. 8.3). Check the calculations of T_H in Fig. 8.4.

(d) Repeat the calculations in question (c) by assuming an elastic cable line.

8.2 Effect of mooring system in the equations of motions

(a) Derive equations (8.26) for the restoring coefficients due to a spread mooring system.

(b) Outline how one can estimate the drag damping from an anchor line. (Hint: Assume the top of the anchorline has a horizontal time dependent motion $\eta_1(t)$. Neglect dynamic effects in the anchorline and find the static position of the anchorline as a function of $\eta_1(t)$. Use the 'cross-flow-principle' to estimate drag-forces along the anchorline, but assume for simplicity that the drag-forces do not influence the anchorline configuration).

8.3 Free-surface effect on propeller

Assume a propeller is not ventilating. The propeller axis is at a submergence of h_0 in still water. The propeller radius is R (see Fig. 8.12). Consider regular waves propagating in the direction of the propeller axis with amplitude ζ_a and circular frequency ω. The wave elevation at the propeller position is

$$\zeta = \zeta_a \sin \omega t$$

Assume the propeller thrust is proportional to the immersed propeller disk area A_1. Set up an expression for the ratio β between the mean thrust in one period T and the thrust for the deeply submerged propeller in terms of h_0, R and ζ_a.

8.4 Turret-moored ship

Consider a turret-moored ship in waves and current. The ship and the coordinate system is defined in Fig. 8.15 and Table 8.1. We will assume the mooring system has zero stiffness in yaw rotation about the turret and that the slowly-varying motion of the ship is a pure yaw motion around the turret.

Assume the incident waves are long-crested and that the ship is in a mean position close to head sea waves. For simplicity we will use the following wave spectrum to describe the sea state

$$S(\omega) = \begin{bmatrix} 3.1 \text{ m}^2 \text{ s} & 0.6 \text{ s}^{-1} < \omega < 1.1 \text{ s}^{-1} \\ 0 & \text{other } \omega\text{-values} \end{bmatrix}$$

The transverse wave drift force F_2^{WD} and yaw moments F_6^{WD} in regular waves with circular frequencies $\omega < 1.1\,\mathrm{s}^{-1}$ are assumed to be frequency-independent. Further we assume F_2^{WD} and F_6^{WD} are proportional to the heading angle β when $\beta < 0.175\,\mathrm{rad}$ where β is defined in

Table 8.1. *Relevant data for the ship to be analysed in exercise 8.4*

Length	$L = 230\,\mathrm{m}$
Beam	$B = 41\,\mathrm{m}$
Draught	$D = 15\,\mathrm{m}$
Displacement	$\nabla = 121\,000\,\mathrm{m}^3$
Turret position: (coordinate system defined in (Fig. 8.15)	$x_T = -20\,\mathrm{m}$
Radii of gyration of the ship mass	$r_{44} = 12\,\mathrm{m}$ $r_{55} = 57.5\,\mathrm{m}$ $r_{66} = 57.5\,\mathrm{m}$
Restoring coefficients with respect to coordinate system in Fig. 8.15	$C_{11} = C_{22} = 3.5 \cdot 10^5\,\mathrm{N\,m}^{-1}$ $C_{26} = 7.0 \cdot 10^6\,\mathrm{N\,rad}^{-1}$ $C_{66} = 1.4 \cdot 10^8\,\mathrm{Nm\,rad}^{-1}$
Added mass, $\omega \to 0$ (with respect to coordinate system in Fig. 8.15)	$A_{11} = 1.02 \cdot 10^7\,\mathrm{kg}$ $A_{22} = 1.29 \cdot 10^8\,\mathrm{kg}$ $A_{26} = 0.0$ $A_{66} = 3.70 \cdot 10^{11}\,\mathrm{kg\,m}^2$

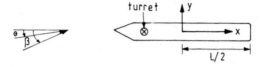

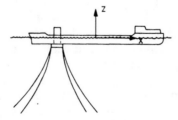

Fig. 8.15. Turret-moored ship (β = wave heading, θ = current angle). Used in exercise (8.4).

Fig. 8.15. We write

$$\frac{F_2^{WD}}{\zeta_a^2} = 0.314 \, \rho g L \beta$$

$$\frac{F_6^{WD}}{\zeta_a^2} = -0.0114 \, \rho g L^2 \beta$$

(ζ_a = wave amplitude of regular incident waves)
The yaw moment is with respect to the z-axis (see Fig. 8.15). The current velocity U_c is $0.5 \, \mathrm{m \, s^{-1}}$. θ is the angle between current and ship's x-axis. The angle between wave propagation direction and current is $10°$ (0.175 rad), i.e.

$$\beta = \theta + 0.175$$

The transverse current force F_2^C and yaw moment F_6^C are written as

$$F_2^C = 0.22 \, \rho U_c^2 L D \theta$$

$$F_6^C = 0.048 \, \rho U_c^2 L^2 D \theta$$

for small angles θ. The yaw moment is with respect to the z-axis.

(a) Find the average position of the ship relative to the wave heading (Answer: $\theta = -0.12$ rad.)

(b) Assume that thrusters are used for heading control and that the average current angle θ is zero. The yaw moment of the thrusters with respect to the turret can be written as in equation (8.40). The natural period of the yaw motion is 400 s. Set up an uncoupled equation for the slowly-varying yaw motion and explain the terms in the equation of motion. How large must the yaw damping effect of the thrusters be in order for the standard deviation of the yaw motion to be less than 3°? (Hint: Use equation (5.48) as a rough approximation.)

8.5 Design data for thruster systems

A vessel with length 185 m and draught 13 m is proposed as a pipelaying vessel. The average wetted area of the ship is $10\,000 \, \mathrm{m^2}$. The vessel should be able to operate up to a significant wave height $H_{\frac{1}{3}} = 4$ m, a mean wave period $T_2 = 7$ s, a wind speed of 26 knots and a current of 2 knots. The vessel will be equipped with a dynamic-positioning system.

(a) Study two conditions. One condition is irregular long-crested head sea waves with current and wind in the longitudinal direction. Another condition is long-crested beam sea waves with current and wind in the transverse direction. Find the mean environmental loads that the thrusters should counteract in the two weather conditions.

(Hint: Estimate wave drift forces by using data given in Fig. 5.13. Estimate current loads by using equation (6.24) and data given in Fig. 6.11. Estimate wind loads by pressure drag formulas. Use drag coefficients equal to 0.8 both for longitudinal and transverse wind directions. The projected areas for longitudinal and transverse wind forces are respectively 350 m² and 1400 m².)

(b) Discuss the possibility that the thrusters lose efficiency due to wave effects by making reasonable choices of thruster positions and estimates of relative vertical motions between the ship and the waves.

9 WATER IMPACT AND ENTRY

SLAMMING

Impulse loads with high pressure peaks occur during impact between a body and water. This is often called 'slamming' and occurs for instance when a ship bottom hits the water with a high velocity. The probability of slamming is highest on the fore part of a ship where the relative vertical velocity between the ship and the waves is largest. Slamming on the ship bottom occurs more often in the ballast condition than in the full-loaded condition and is a larger problem for ships with large block coefficients than for fine ship forms. Wave impacts can also cause bow damage above the waterline. Ship masters normally reduce the speed of a vessel to avoid slamming. An often used criterion for 'voluntary speed reduction' is that a typical ship master reduces the speed if slams occur for more than three out of 100 waves that pass the ship.

For a catamaran, slamming can happen on the underside of the deck between the two hulls. To avoid slamming on platform decks one requires an air gap of typically 1–2 m between the underside of a deck and the most probable highest position the wave can reach relative to the underside of a deck in a '100 year' design condition. The columns of a platform have to be designed for impact loads due to breaking waves (see Fig. 9.1). Inside ship tanks violent fluid motion ('sloshing') can cause high slamming pressures and damages. Abramson et al. (1974) reported that 24 atmospheres of impact pressure had been measured in a tank of an OBO-ship.

The duration of slamming pressure measured at one place on the structure is of the order of milliseconds. It is very localized in space. The position where high slamming pressures occur changes with time. Slamming pressures are sensitive to how the water hits the structure. We can exemplify this with experimental results of slamming pressures due to sloshing in tanks. Abramson et al. (1974) reported model test results for a typical LNG tankform, similar to the one presented in Fig. 1.2. The tank was forced to oscillate harmonically in sway and two-dimensional flow conditions were enforced. Experiments show that even under harmonic oscillations, the pressure variation is neither harmonic nor periodic since the magnitude and duration of the pressure peaks vary from cycle to cycle. A typical histogram for the distribution of peaks is

Fig. 9.1. High slamming pressures can occur when breaking waves hit the column of a platform. The top photo shows a breaking wave approaching the column of a platform and the other photo shows the situation just after the breaking wave has hit the platform. The photos are from the TCP-2 platform at the Frigg Field on the Norwegian Continental Shelf operated by Elf Aquitaine Norge A/S. (Kjeldsen & Andersen (unpublished).)

shown in Fig. 9.2. The most frequently occurring pressure peaks will reach 0.4 times the pressure level exceeded by 10% of all peaks. The 1% exceedance limit is two to three times the 10% exceedance limit. We may also note from Fig. 9.2 that the slamming pressure has a much shorter rise time than decay time.

In order to describe the physics of the slamming problem, let us examine a related, but simpler problem – a horizontal cylinder that is forced through an initially calm water surface with constant velocity V. We assume the body has small submergence, the body has a blunt form and the flow is two-dimensional (see Fig. 9.3). The submergence of the lowest point on the body relative to the calm waterplane is Vt, where t is the time variable. At this stage we will not precisely define how large the wetted area of the body is. This depends on the theoretical approximation that will be used. We will just state that the wetted body area is

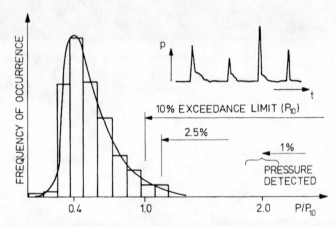

Fig. 9.2. Probability density function of impact pressure peaks from forced harmonic sway oscillation of a LNG-tank. Based on model tests with two-dimensional flow. Also shown is an example of pressure recording. (Abramson *et al.*, 1974.)

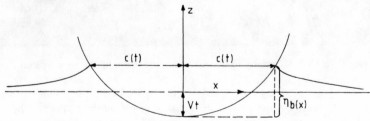

Fig. 9.3. Definition of parameters in analysis of impact forces and pressures on a body.

between $-c(t) \leqslant x \leqslant c(t)$ (see Fig. 9.3). However, we exclude the wetted area due to spray. The reason is that we are interested in finding the hydrodynamic pressure and force on the body and the pressure on the body in the spray area will be very close to atmospheric pressure.

In solving the problem we will assume irrotational flow and an incompressible fluid. This means we can use potential theory. However, in special cases the compressibility of the fluid has to be accounted for. The pressure is assumed constant and equal to atmospheric pressure on the free-surface. However, this is not true when a body with a horizontal flat bottom hits the free-surface. A compressible air pocket is then created between the body and the free-surface in an initial phase (see Fig. 9.4).

To analyse the problem we will use the free-surface condition $\phi = 0$ on $z = 0$. This means we assume the fluid accelerations are much larger than the gravitational acceleration g. The cylinder is replaced by an 'equivalent' flat plate of half-width $c(t)$ in the mean free-surface. The body boundary condition is transferred to the flat plate. We can show this by starting out with the body boundary condition

$$\frac{\partial \phi}{\partial n} \equiv n_1 \frac{\partial \phi}{\partial x} + n_3 \frac{\partial \phi}{\partial z} = -Vn_3$$

where $\mathbf{n} = (n_1, 0, n_3)$ is the normal vector to the body surface. (Positive normal direction is into the fluid.) By a Taylor expansion we can write

$$\frac{\partial \phi}{\partial z} = \frac{\partial \phi}{\partial z}\bigg|_{z=0} - Vt \frac{\partial^2 \phi}{\partial z^2}\bigg|_{z=0} + O(V^2 t^2)$$

Further, for slamming to have any practical meaning $n_1 \ll n_3$ and $n_3 \approx -1$. This means the boundary condition is approximately

$$\frac{\partial \phi}{\partial z} = -V \quad \text{on} \quad z = 0$$

when $Vt/c(t)$ is small. The boundary-value problem that we have to

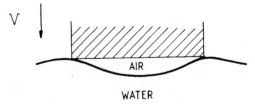

WATER

Fig. 9.4. Deformation of the free-surface and formation of an air pocket during entry of a body with a horizontal flat bottom.

solve at every time instant is illustrated in Fig. 9.5. A solution to this problem may be found in many text-books (see for instance Newman, 1977 on p. 122). We can write the velocity potential on the body as

$$\phi = -V(c^2 - x^2)^{\frac{1}{2}}, \qquad |x| < c(t) \tag{9.1}$$

The pressure on the body follows from Bernoulli's equation. However, we have to be consistent. In formulating the body boundary condition we neglected terms because $Vt/c(t)$ is small. The same has to be done when evaluating the pressure p. The pressure can be written as

$$p = -\rho \frac{\partial \phi}{\partial t} - \rho g z - \frac{\rho}{2} \mathbf{V} \cdot \mathbf{V} + C$$

where C is a constant. The slamming occurs over a small time instant. This means the rate of change of ϕ with time is generally larger than the rate of change of ϕ with respect to x and z. We may therefore neglect

$$-\frac{\rho}{2} \mathbf{V} \cdot \mathbf{V} = -\frac{\rho}{2} \left(\left(\frac{\partial \phi}{\partial x}\right)^2 + \left(\frac{\partial \phi}{\partial z}\right)^2 \right)$$

relative to $-\rho \, \partial \phi / \partial t$. Further, since the submergence of the body is small we may also neglect the hydrostatic pressure term $-\rho g z$ relative to $-\rho \, \partial \phi / \partial t$. We may therefore approximately write the hydrodynamic pressure

$$p = -\rho \frac{\partial \phi}{\partial t} = \rho V \frac{c}{(c^2 - x^2)^{\frac{1}{2}}} \frac{dc}{dt} \tag{9.2}$$

The corresponding vertical force on the body is

$$F_3 = \int_{-c}^{c} p \, dx = \rho V c \frac{dc}{dt} \int_{-c}^{c} \frac{dx}{(c^2 - x^2)^{\frac{1}{2}}} = V \frac{d}{dt} \left(\rho \frac{\pi}{2} c^2 \right) \tag{9.3}$$

We note that $\rho(\pi/2)c^2$ is the added mass in heave of the flat plate when $\omega \to \infty$. This is half the added mass for the flat plate in infinite fluid. We will now illustrate how we can find the wetted body length ($\approx 2c(t)$) for a circular cylinder and a wedge.

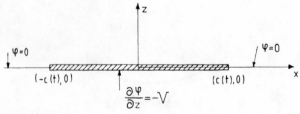

Fig. 9.5. Boundary value problem in simplified analysis of impact between a two-dimensional body and water.

Case 1. Circular cylinder

We will assume the wetted length can be measured from the calm water level. By geometry it follows that

$$c^2(t) = 2VtR - V^2t^2 \tag{9.4}$$

where R is the cylinder radius. By using equations (9.3) and (9.4) and a definition of a slamming coefficient C_s, it follows that

$$C_s \equiv \frac{F_3}{\frac{1}{2}\rho V^2 2R} = \pi \tag{9.5}$$

at the initial time of impact. It is well known that C_s is larger in experiments. For instance, Campbell & Weynberg's (1980) experimental value of C_s is 5.15 at the time of impact. If equation (9.3) for the force is correct, it indicates that the wetted length should be larger. We will analyse this by following Wagner's (1932) approach. The previous approach is referred to as von Karman's (1929) solution.

The first step is to find an expression for the free-surface elevation. From the solution of the boundary-value problem illustrated in Fig. 9.5 it follows that

$$\frac{\partial \phi}{\partial z} = \frac{Vx}{\sqrt{[x^2 - c^2(t)]}} - V \quad \text{at} \quad z = 0, \ x > c(t) \tag{9.6}$$

The free-surface elevation η relative to the bottom of the cylinder can then be written as

$$\eta = \int_0^t \frac{Vx}{\sqrt{[x^2 - c^2(t)]}} \, dt \tag{9.7}$$

This has to be equal to $\eta_b(x)$ which is the vertical coordinate of a point on the cylinder relative to the bottom of the cylinder (see Fig. 9.3). This means

$$\eta_b(x) = \int_0^x \frac{x\mu(c)}{\sqrt{[x^2 - c^2(t)]}} \, dc \tag{9.8}$$

where

$$\mu(c) = V \, dt/dc \tag{9.9}$$

We do not know $\mu(c)$. Equation (9.8) is therefore an integral equation that determines $\mu(c)$. When $\mu(c)$ is found we can use equation (9.9) to find c as a function of time. We will try to find an approximate solution of equation (9.8) by guessing that

$$\mu(c) \approx A_0 + A_1 c \tag{9.10}$$

Here A_0 and A_1 are unknown constants. By integrating the right hand side of equation (9.8) it follows that

$$\eta_b(x) = A_0 \frac{\pi}{2} x + A_1 x^2 \tag{9.11}$$

For a circular cylinder it follows that

$$x^2 + (R - \eta_b)^2 = R^2$$

i.e.

$$x^2 \approx 2\eta_b R$$

This means $A_0 = 0$ and $A_1 = 1/(2R)$. By using equation (9.9) it follows that

$$Vt = \int_0^c \frac{c}{2R} \, dc$$

This means

$$c = 2\sqrt{(VtR)} \tag{9.12}$$

By comparing equation (9.12) with equation (9.4) we see that the wetted length is $\sqrt{2}$ times the wetted length measured from the calm water. The slamming coefficient C_s is found to be 2π at the initial time of impact. This is higher than the experimental value of 5.15 found by Campbell & Weynberg (1980).

The hydrodynamic pressure on the body can partly be obtained by equation (9.2). The pressure coefficient C_p is given by

$$C_p = \frac{p}{\frac{1}{2}\rho V^2} = \frac{4}{\sqrt{\left[4\left(\frac{Vt}{R}\right) - \left(\frac{x}{R}\right)^2\right]}} \quad \text{for} \quad |x| < c(t) = 2\sqrt{(VtR)}$$

$$\tag{9.13}$$

The result can be compared with the experimental values by Campbell & Weynberg (see Fig. 9.6). The agreement is reasonable except for the maximum pressure value. According to equation (9.13) $C_p \to \infty$ when $x \to c(t)$. This is partly a consequence of incorrect boundary conditions on the free surface in the immediate vicinity of $|x| = c(t)$. If we use a microscopic view of the flow in this area we will find the fluid moves with a very high velocity (see Fig. 9.7). This can be seen from the velocity $dc/dt = (VR/t)^{\frac{1}{2}}$ of the wetted area. In the jet flow which is present at $x > c(t)$, the pressure is nearly constant across the jet. This means the pressure on the body is close to atmospheric pressure. The maximum hydrodynamic pressure in the immediate vicinity of $|x| = c(t)$ is according to Wagner (1932) and Armand & Cointe (1986) equal to $0.5\rho(dc/dt)^2$ (see Fig. 9.7), i.e.

$$C_{p_{max}} = (dc/dt)^2/V^2 \tag{9.14}$$

Using Wagner's solution this means that $C_{p_{max}} = R/(Vt)$ which occurs at $|x| = 2(VtR)^{\frac{1}{2}}$. When for instance $Vt/R = 0.012$ it means that $C_{p_{max}} = 83$. This is about twice the measured value by Campbell & Weynberg (1980). However, the position where the maximum pressure occurs is approximately correct. On the other hand experimental errors are possible, one reason being that the slamming peak pressure depends strongly on the area over which the pressure is measured.

If we used von Karman's solution then $C_{p_{max}} = 0.5R/(Vt)$ at

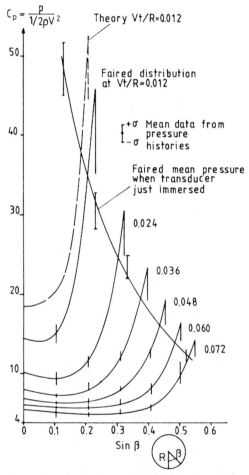

Fig. 9.6. Comparison between experimental and theoretical values of slamming pressure p during entry of a circular cylinder with constant downward velocity V (t = time variable with $t = 0$ corresponding to initial time of impact). Experiments: Campbell & Weynberg (1980). (Adapted from Campbell & Weynberg, 1980.)

$|x| = (2VtR)^{\frac{1}{2}}$. This is in satisfactory agreement with the maximum pressure value at $Vt/R = 0.012$, but it occurs at the wrong position on the cylinder. If we plot $C_{p_{max}}$ as a function of time we see that it goes to infinity when $t \to 0$. One important reason for this is that we have neglected compressibility effects in the water. When these are accounted for the pressure cannot be higher than

$$p_{ac} = \rho c_e V \qquad (9.15)$$

This is called the acoustic pressure where c_e is the velocity of sound in

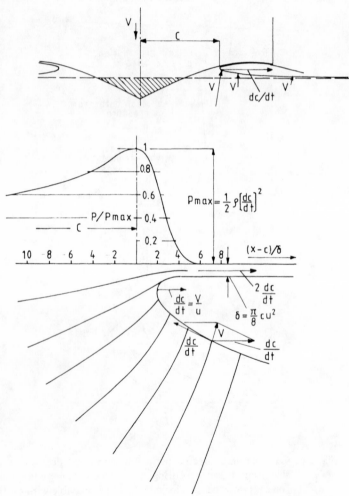

Fig. 9.7. Details of the flow at the intersection between the free surface and a body during impact. (p = hydrodynamic pressure.) (Wagner, 1932.)

the water. With no air content c_e varies typically between 1450 m s^{-1} and 1540 m s^{-1}. In rough seas, air bubbles will be in the water altering the speed of sound dramatically (Lundgren, 1969). We can show equation (9.15) by a simple analysis. Let us assume a body hits the water with a horizontal flat surface A. We neglect all free-surface deformation and effects of the air. At the time of contact between the body and the free-surface, the fluid will be given a disturbance that will propagate with the sound velocity c_e. During the time step Δt after the impact a mass of fluid $M \approx \rho c_e A \, \Delta t$ will be accelerated. The acceleration is approximately $V/\Delta t$. This means there has to be a force on the body that is equal to $\rho c_e A \, \Delta t V / \Delta t$. The average pressure on the body is therefore $\rho c_e V$.

The importance of the acoustic pressure is illustrated in Fig. 9.8 where pressure results from forcing a vertical cylinder with radius 5 m through the free-surface with velocity $V = 8$ m s^{-1} are presented. We note that the acoustic pressure is about 125 times the atmospheric pressure and is the maximum pressure occurring close to the centre line of the cylinder. When $x > \approx 0.5$ m the slamming pressure is determined from Wagner's analysis. As pointed out earlier the slamming pressures are very local in

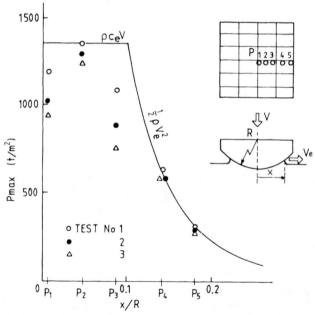

Fig. 9.8. Maximum impact pressures on a circular cylinder as a function of time after initial impact ($V = 8$ m s^{-1}, $R = 5$ m). (Hagiwara & Yuhara, 1976.)

time and space and the maximum slamming pressure does not occur simultaneously around the cylinder surface.

In design of local structural parts against slamming loads, the peak value of the slamming pressure gives a conservative estimate of the load distribution. This has been discussed by Hagiwara & Yuhara (1976). They introduce an 'equivalent static pressure' p_{eq} in analysing the strain of a rectangular panel due to slamming load. p_{eq} is defined as uniformly distributed static pressure which causes static strain equal to the maximum transient strain during impact. In an example with a 0.45 m × 0.31 m panel, which is part of a circular cylinder surface of radius 5 m, they found that the magnitude of p_{eq} was about one-third of the maximum impact pressure p_{max}. However, this result will obviously be a function of the size of the panel.

Case 2. Wedge

We will apply Wagner's (1932) analysis to a wedge. We define a deadrise angle β which is the angle between the wedge side and a horizontal line through the wedge apex. The vertical coordinate $\eta_b(x)$ of the wedge surface relative to the wedge apex can be written as

$$\eta_b(x) = x \tan \beta \tag{9.16}$$

It now follows from equation (9.11) that $A_0(\pi/2) = \tan \beta$. From equations (9.10) and (9.9) it follows that

$$c(t) = \frac{\pi V t}{2 \tan \beta} \tag{9.17}$$

From equation (9.14) it follows that

$$C_{p_{max}} = \frac{\pi^2}{4} \cotan^2 \beta \tag{9.18}$$

Theoretical calculations are shown in Fig. 9.9 together with experimental results. The calculations are in fair agreement with experimental results. We note that equation (9.18) breaks down when $\beta \rightarrow 0$, i.e. when the bottom of the wedge becomes more and more flat. When a flat-bottomed body enters the free-surface, it is known that the airflow between the body and the free-surface is important at time instances when the airgap between the body surface and the free-surface is small relative to the transverse dimensions of the body (Verhagen, 1967). The free-surface elevation depends on the pressure in the air above the free-surface and the pressure in the air depends on the gap between the free-surface and the bottom. The water elevation is first notice-able at the edges of the body (see Fig. 9.4). This causes the air to be entrapped between the free-surface and the body. If the body has a small deadrise angle, say 2–3°, the air will not be entrapped (Koehler &

Kettleborough, 1977). The entrapped air will have a pronounced cushioning effect, and the impact pressure on the hull is therefore sensitive to small changes in angle between bottom and water surface. This indicates that it is difficult to create deterministic flow situations for impact problems. This was also pointed out in the discussion of Fig. 9.2.

When calculating bottom slamming on ships the relative vertical velocity between the ship and the waves is an important quantity. We ought to be precise about what we mean by vertical velocity. It is not the vertical velocity in a (x, y, z) system fixed in space (see Fig. 3.13), but it is the z''-component of the velocity in a coordinate system (x'', y'', z'') that is fixed relative to the ship and coincides with the (x, y, z)-system when the ship does not oscillate (see Fig. 3.13).

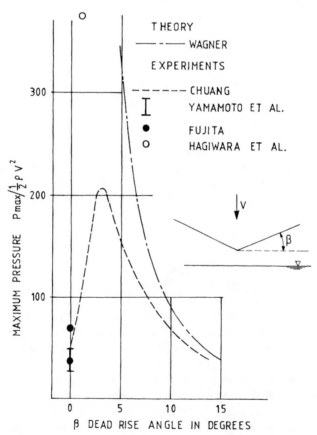

Fig. 9.9. Variation of maximum impact pressure with deadrise angle for a wedge shaped body. (Hagiwara & Yuhara, 1976.)

We have also drawn a picture of the ship's velocity U in the figure. According to an observer on a ship, the ship's speed acts as an incident constant velocity U. From the figure we see that this velocity has a z''-component that is equal to $U\eta_5$. This has to be accounted for in the calculation of the z''-component of the relative velocity. From this discussion we can write up a linear z''-component of the relative velocity as

$$V_R = \frac{d\eta_3}{dt} - x\frac{d\eta_5}{dt} - U\eta_5 - w \tag{9.19}$$

where w is the z-component of the undisturbed wave velocity in the free-surface at the longitudinal position x where ship slamming is to be analysed. The heave and pitch motions may, for instance, be obtained by a strip theory program if regular incident waves are analysed.

Statistical estimates of slamming pressure in an irregular sea can easily be obtained if we assume that the magnitude of the impact pressure can be written as

$$p = \tfrac{1}{2}\rho k \, |V_R|^2 \tag{9.20}$$

Here k depends on the sectional form, especially the local deadrise angle at the point considered (see for instance Fig. 9.9). We will assume short-term stationarity of the irregular waves, i.e the waves can be described by a sea spectrum $S(\omega)$. For simplicity we will assume the waves to be long-crested. If we assume that the maxima (peak values) of the relative velocity of the ship follow a Rayleigh distribution, we can write the probability that the maximum of the relative velocity is larger than v as

$$P(|V_R| > v) = \exp\left(-\frac{v^2}{2\sigma_v^2}\right) \tag{9.21}$$

where

$$\sigma_v^2 = \int_0^\infty S(\omega) \left|\frac{V_R(\omega)}{\zeta_a}\right|^2 d\omega \tag{9.22}$$

is the variance of the relative velocity and $|V_R(\omega)/\zeta_a|$ is the transfer function of the relative velocity. The latter can be obtained by analysis of the ship response in incident regular waves of circular frequency ω with the same wave heading as the irregular sea. By combining the pitch and heave velocities (see equation (9.19)) we can find the amplitude of the relative velocity $|V_R(\omega)|$ divided by the incident wave amplitude ζ_a. The results are ship speed and wave heading dependent. The relative velocity is normally largest on the fore part of the ship, but large relative velocities may also cause slamming problems on exposed aft parts of the ship. In order for slamming to occur at a point A on the surface of the ship it is necessary that the relative vertical motion at the same longitudinal position of the ship is larger than the vertical distance d from the still water surface to point A. The probability for the amplitude of the relative motion η_R to be larger than d can be expressed as

$$P(\eta_R > d) = \exp\left(-\frac{d^2}{2\sigma_r^2}\right) \tag{9.23}$$

where

$$\sigma_r^2 = \int_0^\infty S(\omega) \left|\frac{\eta_R(\omega)}{\zeta_a}\right|^2 d\omega \tag{9.24}$$

is the variance of the relative motion. This expression does not account for the waves being influenced by the ship. In the calculation of the relative motion we should take this into account.

Using the fact that the relative motion and relative velocity are statistically independent, we can write the probability that the slamming pressure becomes larger than a given value p at a specific point A on the ship as

$$P(\text{impact pressure} > p) = \exp\left[-\left(\frac{p}{\rho k \sigma_v^2} + \frac{d^2}{2\sigma_r^2}\right)\right] \tag{9.25}$$

Equation (9.25) has been obtained by multiplying together equations (9.23) and (9.21), and then using equation (9.20) with $|V_R| = v$ to rewrite equation (9.21).

From equation (9.25) we find the most probable largest slamming pressure p_{max} in N encountered waves to be

$$p_{max} = \rho k \sigma_v^2 \left(\log N - \frac{d^2}{2\sigma_r^2}\right) \tag{9.26}$$

where N can be approximated by t/T_2. Here t is the time duration of the N encounter waves and T_2 is the mean wave period. Both σ_v, σ_r and k depend on the position of the ship where slamming is to be evaluated. It should be kept in mind that equation (9.26) is based on using equation (9.20), i.e. no compressibility effects are accounted for (see equation (9.15)). We could have substituted equation (9.15) into equation (9.21) by setting $V = v$. In this way we would find the most probable largest acoustic pressure p_{max}^A and use the smallest of p_{max} and p_{max}^A.

Long-term predictions of slamming pressure can be obtained by combining equation (9.25) with the joint probability of significant wave height $H_{\frac{1}{3}}$ and mean wave period T_2. This is similar to the method described in chapter 3 for long-term predictions of linear wave-induced motions and loads.

In a given sea state we can also use equation (9.25) to evaluate the probability of slamming. It is then necessary to define how large a pressure has to be to call it slamming. One indirect way is to define a threshold velocity V_{cr} and say that slamming occurs when the relative velocity is larger than V_{cr}. For instance Ochi (1964) sets the threshold velocity equal to

$$V_{cr} = 0.093(gL)^{\frac{1}{2}} \qquad (9.27)$$

where L is the ship length. The probability that slamming occurs can then be written as

$$P(\text{slamming}) = \exp\left(-\left(\frac{V_{cr}^2}{2\sigma_v^2} + \frac{d^2}{2\sigma_r^2}\right)\right) \qquad (9.28)$$

This can for instance be used for studies of voluntary speed reduction. An often used criterion is that a typical ship master reduces the speed if slams occur more than 3 of 100 times that waves pass the ship. This means that if P(slamming) is calculated to be larger than 0.03, the ship speed U used in evaluating σ_v and σ_r in equation (9.28) is too high. It should be reduced to a level so that P(slamming) is less than 0.03. However, it should be kept in mind that additional criteria are also used for 'voluntary' speed reduction. Ochi & Motter (1974) say that for a ship in fully-loaded condition voluntary speed reduction does *not* occur if

$$P(\text{water on deck}) \leqslant 0.07$$

and (or)

$$P(\text{significant acceleration in the bow} > 0.4\,\text{g}) < 0.07$$

For a ship in *ballast* condition voluntary speed reduction does not occur if

$$P(\text{slamming}) \leqslant 0.03$$

and (or)

$$P(\text{significant acceleration in the bow} > 0.4\,\text{g}) \leqslant 0.03$$

Fig. 9.10 shows an example of calculations of voluntary speed reduction for a 170 m long ship in head sea long-crested waves as a function of sea state or significant wave height. For smaller sea states the involuntary speed reduction due to added resistance in waves and wind determines the reduction in ship speed at full engine power.

WATER ENTRY PROBLEMS

In the case of a horizontal jacket truss in the splash zone one is not particularly interested in the high *pressures* occurring when the water hits the member. One is more interested in the *force* on the body as the cylinder hits the water and proceeds through the water. One of the first calculations of this type was done by von Karman (1929) for a related problem with seaplanes. He used the free-surface condition $\phi = 0$ and measured the wetted length from the calm water level. The force was written as $\mathrm{d}(A_{33}V)/\mathrm{d}t$ where A_{33} is the infinite-frequency added mass coefficient in heave for the body as a function of submergence. The added mass calculations were simplified by using flat plate results giving the same result as equation (9.3) when $c(t)$ is measured from the intersection point between the cylinder and the still water level. The formula does not account for buoyancy and incident wave effects.

The formula can be derived by momentum considerations. We will show this when there are no incident wave effects. Infinite water depth is assumed. We start out with equation (5.7) which says that

$$\frac{\mathrm{d}\mathbf{M}}{\mathrm{d}t} = -\rho \iint_{S} \left[\left(\frac{p}{\rho} + gz \right) \mathbf{n} + \mathbf{V}(V_n - U_n) \right] \mathrm{d}s \qquad (9.29)$$

We should note that the positive normal direction is out of the fluid domain. The momentum of the fluid inside S can be written as

$$\mathbf{M}(t) = \iint_{S} \rho \phi \mathbf{n} \, \mathrm{d}s \qquad (9.30)$$

This follows from Gauss' theorem (see equation (5.6)). By using $\phi = 0$ on the still water level and neglecting free-surface waves, we get no contribution from the free-surface integration in equations (9.29) and (9.30). Further, by replacing the pressure p at S_∞ by Bernoulli's

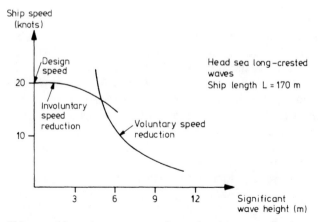

Fig. 9.10. Ship speed in various sea states due to involuntary and voluntary speed reduction.

equation, it follows that

$$\mathbf{F} = -\rho \frac{\mathrm{d}}{\mathrm{d}t} \iint_{S_B} \phi \mathbf{n} \, \mathrm{d}s - \iint_{S_B} \rho g z \mathbf{n} \, \mathrm{d}s$$

$$- \rho \iint_{S_\infty} [\mathbf{V} V_n - \tfrac{1}{2} |\mathbf{V}|^2 \mathbf{n}] \, \mathrm{d}s \tag{9.31}$$

It is important to note that $\mathrm{d}/\mathrm{d}t$ in equation (9.31) is an ordinary time derivative. The integral over S_∞ in equation (9.31) can be neglected. This follows by studying how the disturbance from the body dies out far away from the body. One way to find this out is by representing the flow by a distribution of sources (and sinks) over the wetted body surface (see chapter 4). This means we write

$$\phi(x, y, z) = \iint_{S_B} Q(s)[((x - \xi(s))^2 + (y - \eta(s))^2 + (z - \zeta(s))^2)^{-\frac{1}{2}}$$

$$- ((x - \xi(s))^2 + (y - \eta(s))^2 + (z + \zeta(s))^2)^{-\frac{1}{2}}] \, \mathrm{d}s \tag{9.32}$$

The term in the brackets satisfies the free-surface condition $\phi = 0$ on $z = 0$. The source density $Q(s)$ is determined by satisfying the body boundary condition. If we let $r = (x^2 + y^2 + z^2)^{\frac{1}{2}}$ be large we can simplify the term in the brackets. By considering it as a function of $(\xi(s), \eta(s), \zeta(s))$ and making a Taylor expansion about $(0, 0, 0)$ we find the dominant term to be

$$-2\zeta z r^{-3} \tag{9.33}$$

This means the term in the brackets of equation (9.32) behaves like a dipole with a vertical axis. By differentiating (9.32) with respect to x, y and z we see that the velocity due to the body decays far away at least as fast as r^{-3}. Going now back to equation (9.31) it means that the order of magnitude of the integrand of the integral over S_∞ is r^{-6}. Since the surface element $\mathrm{d}s$ is proportional to r^2, it means that the integral over S_∞ is zero in equation (9.31).

The first term in equation (9.31) can be written in terms of added mass. We will illustrate this by limiting ourselves to studying the vertical force due to vertical motion. This means the velocity potential ϕ satisfies the body boundary condition

$$\frac{\partial \phi}{\partial n} = -V n_3 \quad \text{on} \quad S_B \tag{9.34}$$

In addition the velocity potential satisfies the free-surface condition

$$\phi = 0 \quad \text{on} \quad z = 0 \tag{9.35}$$

For a general body this problem can be solved as outlined by equation (9.32). This is the same boundary-value problem we solved to find the added mass in heave when the frequency of oscillation $\omega \to \infty$. This means we can use results from that problem and write

$$\rho \iint_{S_B} \phi n_3 \, ds = -V A_{33} \tag{9.36}$$

The second term in equation (9.31) is the buoyancy term. This means we have shown that the hydrodynamic (including hydrostatic) vertical force on a body penetrating the free-surface with a downward velocity V can be written as

$$F_3 = \frac{d}{dt}(A_{33}V) + \rho g \Omega \tag{9.37}$$

where Ω is submerged volume and A_{33} is the high-frequency added mass in heave for the body as a function of submergence relative to calm water level. The same result applies in two dimensions. Fig. 9.11 shows some results for an infinitely long horizontal circular cylinder. Two-dimensional added mass values in heave $A_{33}^{(2D)}$ are shown as a function of h/R. R is the cylinder radius and h is the submergence as defined in the figure. The asymptotic value of $A_{33}^{(2D)}/(\rho R^2)$ for large h/R is π. $A_{33}^{(2D)}$ has been calculated numerically by source technique, but can also be calculated by analytical expressions (see the review article by Greenhow & Li 1987). The two force terms in equation (9.37) are also shown in Fig. 9.11 in the case when the forced velocity is constant. We can then write

$$\frac{d}{dt}\{A_{33}^{(2D)}(t)V\} = \frac{dA_{33}^{(2D)}}{dh} V^2 \tag{9.38}$$

and define a C_s-value by equating

$$\frac{dA_{33}^{(2D)}}{dh} V^2 = \frac{\rho}{2} C_s 2R V^2 \tag{9.39}$$

The experimental values by Campbell & Weynberg (1980) for a smooth circular cylinder have also been shown in Fig. 9.11. The experimental values can be represented by the formula

$$C_s = 5.15/(1 + 8.5h/R) + 0.275h/R \tag{9.40}$$

Buoyancy effects have not been deducted from this formula. Campbell & Weynberg estimated the error in equation (9.40) due to buoyancy effects to be from 0.05 to 0.54. This may be one reason why the theoretical values for C_s are higher than the experimental values in the range between $h/R = 1.0$ and 2.0. On the other hand one may question the theoretical assumptions when h/R is close to 2.0 and larger. This can be illustrated by the photos in Fig. 9.12. The upper part of the cylinder is dry even when $h/R > 2.0$. One reason for the differences between experimental and numerical values of C_s is that the present theory does not adequately predict the wetted surface.

Equation (9.37) can be generalized to include incident wave effects. We will assume the wavelength is large relative to the cross-section of the body so that the undisturbed velocity and pressure field can be assumed to have small variation over the submerged volume occupied by the body.

We assume regular sinusoidal deep water waves that can be described by linear wave theory. If the waves propagate along the positive x-axis the incident wave potential is

$$\phi_0 = \frac{g\zeta_a}{\omega} e^{kz} \cos(\omega t - kx) \tag{9.41}$$

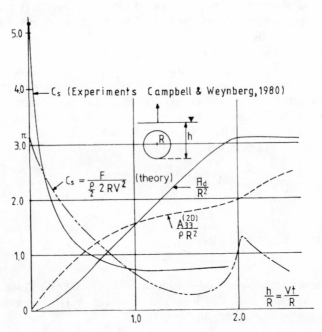

Fig. 9.11. Slamming coefficient C_s, added mass $A_{33}^{(2D)}$ and displaced volume A_d of a circular cylinder as a function of submergence. (F = vertical force, V = constant downward velocity of the cylinder, t = time variable with $t = 0$ corresponding to initial time of impact.)

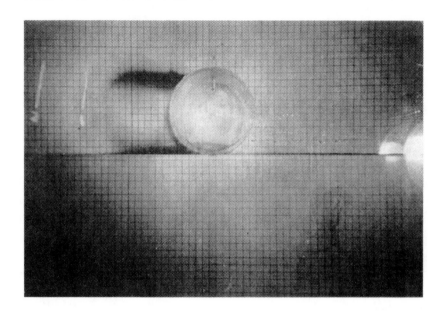

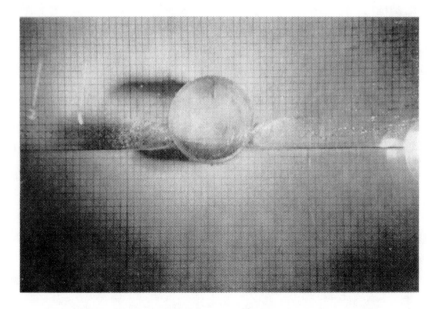

Fig. 9.12. Flow visualizations from impact studies of a circular cylinder.
(Greenhow & Lin, 1983.)

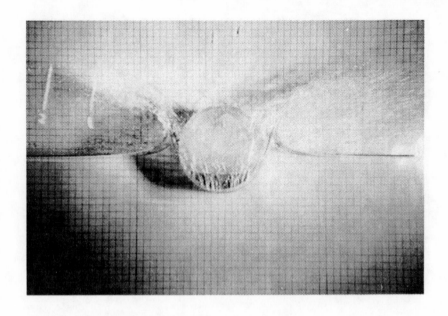

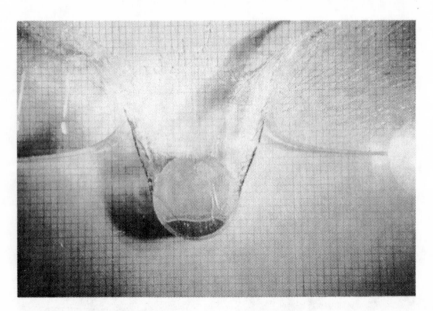

Fig. 9.12. (*continued*)

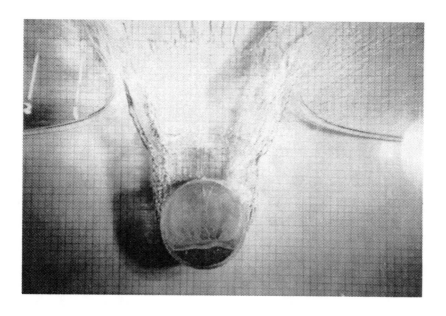

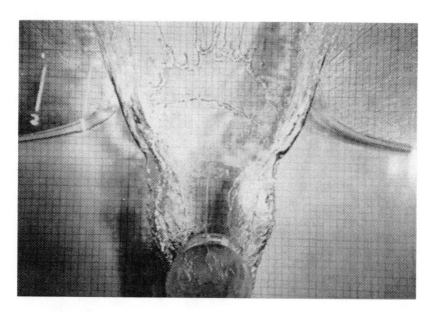

Fig. 9.12. (*continued*)

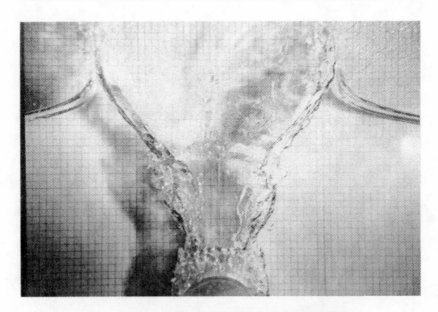

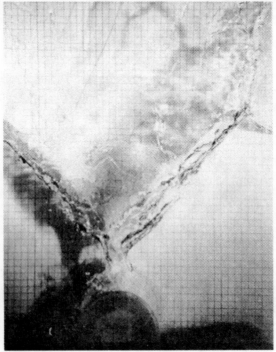

Fig. 9.12. (*continued*)

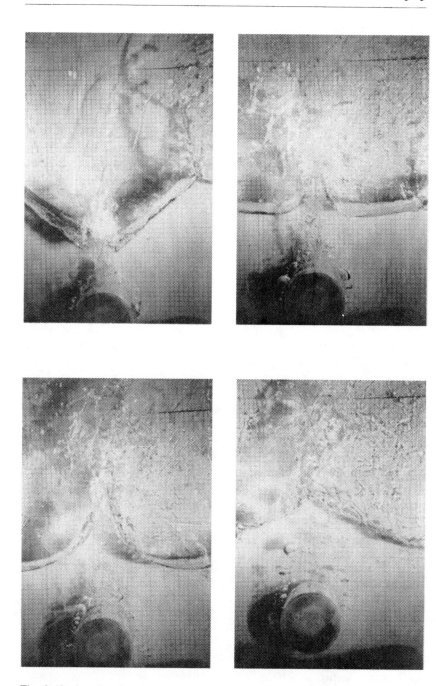

Fig. 9.12. (*continued*)

However, any wave heading can be considered. The flow acceleration generated by the body is assumed to be much larger than the gravitational acceleration in the neighbourhood of the body. The velocity potential ϕ_B due to the body will then approximately satisfy the free-surface condition

$$\phi_B = 0 \tag{9.42}$$

on the instantaneous position $z = \zeta_a \sin(\omega t - kx)$ of the incident wave field. Condition (9.42) is relevant for high-speed water entry problems.

We can proceed in a similar way as for the derivation of equation (9.37). If we limit ourselves to analysing the vertical force F_3 and assume the body has a vertical motion η_3 relative to a coordinate system fixed in space, with positive direction upwards, we can show that

$$F_3 = \rho\Omega(t)\frac{dw}{dt} + \rho g\Omega(t)$$

$$- \frac{d}{dt}\left(A_{33}\left(\frac{d\eta_3}{dt} - w\right)\right) \tag{9.43}$$

where

$$w = \omega\zeta_a \cos(\omega t - kx), \tag{9.44}$$

$\Omega(t)$ is the instantaneous submerged volume of the body, and x is the average x-coordinate of $\Omega(t)$. The two first terms in equation (9.43) are due to the combined effect of the Froude–Kriloff pressure and the hydrostatic pressure.

Equation (9.43) can be used in the analysis of lifting operations of a subsea structure that is lowered from a surface vessel to the sea bed (see Fig. 9.13). During the transit of the structure through the water surface high hydrodynamic forces may occur. This can cause slack in the hoisting wire followed by high snatch loads. This may also happen during other phases of the operation, for instance when the structure is close to the sea bed. On the sea bed the boundary condition $\partial\phi_B/\partial z = 0$ applies. Greenhow & Li (1987) have presented analytical solutions for added mass as a function of distance away from the sea bed (see also Fig. 3.11).

Equation (9.44) can be further generalized to irregular long-crested seas by writing

$$w = \sum_{i=1}^{N} \omega_i A_i \cos(\omega_i t - k_i x + \epsilon_i) \tag{9.45}$$

where A_i is determined from the sea spectrum and ϵ_i are random phase angles (see equation (2.22)). It is necessary that the wavelength of each wave component in (9.45) is large relative to the cross-sectional dimension of the structure. Equation (9.45) can also be generalized to short-crested seas.

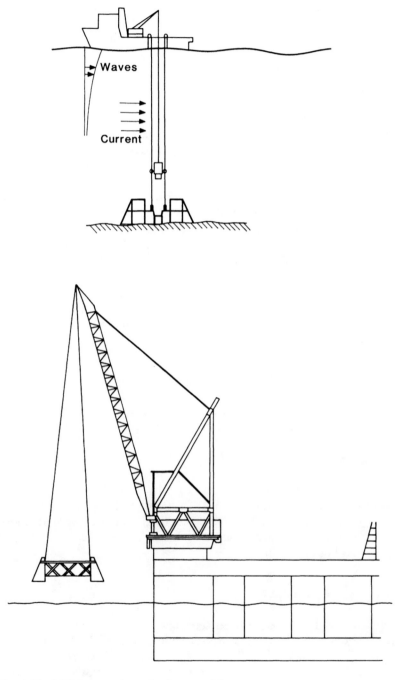

Fig. 9.13. Lifting operations of subsea modules.

To evaluate the hydrodynamic forces we need to know the infinite frequency added mass in heave A_{33} as a function of submergence. This can be calculated independently of the wave field calculation. What matters is the submergence relative to the water level. This is exactly the same type of calculation we presented in Fig. 9.11 for a circular cylinder. Miloh (1981) presents similar results for a sphere.

We should not use equation (9.43) when the structure is fully submerged and close to the free-surface. The reason is that the free-surface condition $\phi_B = 0$ used in calculating ϕ_B is questionable and influences the results. When the body is deeply submerged, we may again use equation (9.43). In this case $dA_{33}/dt = 0$. For the exit problem it is not possible to use equation (9.43). Some exact calculations within potential flow theory with correct non-linear free-surface conditions have been given by Greenhow (1988) and Telste (1987). We should note the very different flow pictures between an exit and an entry of a cylinder. Fig. 9.14 shows an example of flow visualizations of an exit of a cylinder. The water above the cylinder is lifted with the cylinder. In the last picture the free-surface breaks spontaneously under the cylinder.

During the first phase of the impact we may underestimate the forces (see Fig. 9.11). However, what is important for the global response of the structure during a small time increment Δt after the impact is the impulse

$$I = \int_0^{\Delta t} F_3(t) \, dt \tag{9.46}$$

relative to the mass of the structure. A small time increment means a small time relative to a natural period. Typically the interesting natural periods will be 1–2 s. However, this depends on the stiffness properties of the crane. Equation (9.43) is a good approximation in the calculation of the impulse.

Equation (9.43) can be further generalized and applied for a ship at forward speed (see exercise 9.4). An example is the evaluation of hydrodynamic forces on bow flare sections. This can cause large midship stresses with resulting damage to deck and bottom plates. Loss of bow parts has been reported.

EXERCISES

9.1 Impact loads on horizontal bracing

A horizontal member (of a jacket) with diameter $D = 0.5$ m is in the splash zone (Fig. 9.15). Regular sinusoidal waves of period 15 s and height $H = 30$ m propagate in deep water in the positive x-direction. Find the vertical slamming force (i.e. the force when the wave hits the

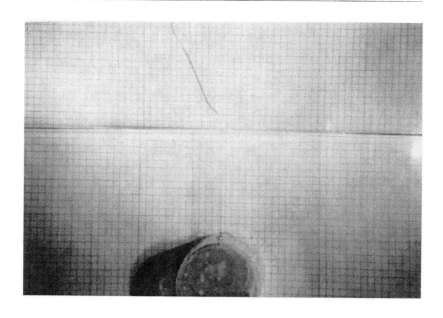

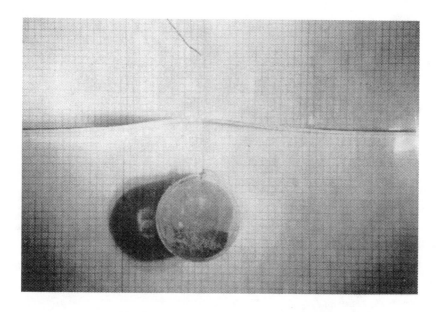

Fig. 9.14. Flow visualizations from exit studies of a circular cylinder.
(Greenhow & Lin, 1983.)

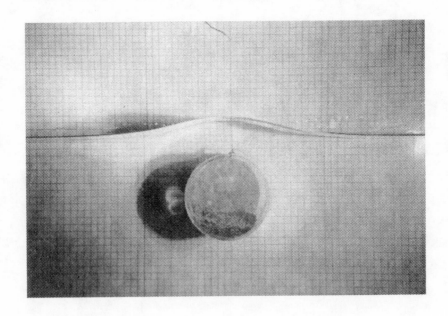

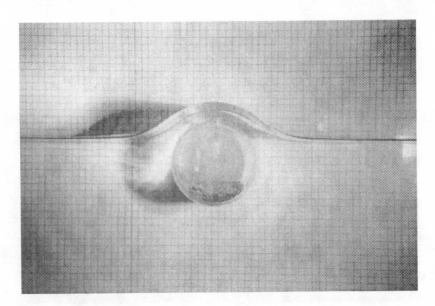

Fig. 9.14. (*continued*)

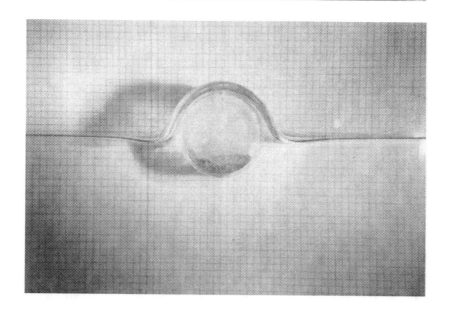

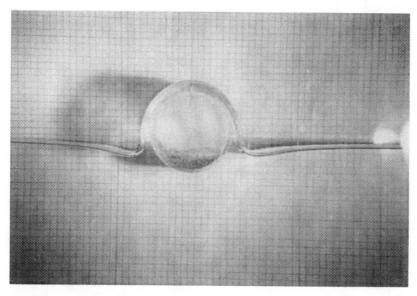

Fig. 9.14. (*continued*)

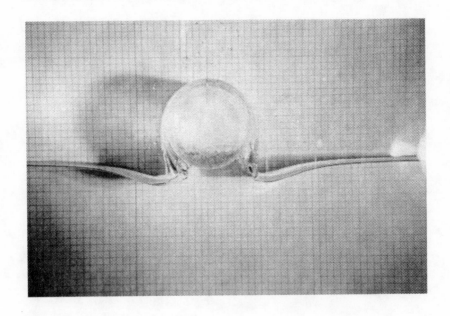

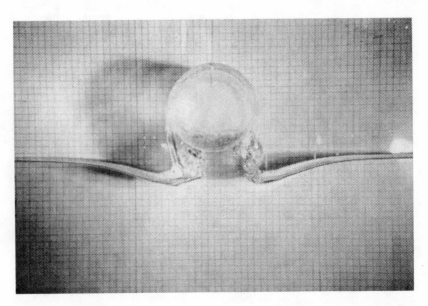

Fig. 9.14. (*continued*)

member) as a function of d. (d is the height of the centre of the member from the still water level).

9.2 Impact loads on a vertical cylinder

Consider a situation where a vertical face of a wave hits a vertical cylinder (see Fig. 9.1). Assume the waves are long-crested before impact (see Fig. 9.16). Set the horizontal velocity of the wave front equal to $(g/(2\pi))T$ (i.e. the phase velocity of small amplitude deep-water waves with period T). Choose $T = 10$ s and cylinder radius 10 m.

Calculate the impact pressure as a function of time and space.

9.3 Lifting operations

Consider a watertight circular cylinder of length $L = 10$ m and diameter $D = 2$ m that is lowered horizontally from a crane ship (see Fig. 9.17). The mass of the cylinder is $\pi \cdot 10^4$ kg. The stiffness of the crane-wire system is 2000 kN m^{-1} when the cylinder is given a forced displacement in the vertical direction. When the cylinder goes through the water surface, the waves hit the cylinder with a vertical velocity 2 m s^{-1}. Assume the water surface is horizontal at the time of impact and that the cylinder and the crane top have no vertical motion before impact.

Assume the vertical motion η_3 of the cylinder can be described by a

Fig. 9.15. Horizontal bracing in the wave zone.

Fig. 9.16. Impact pressure from waves breaking against a vertical cylinder.

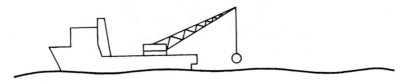

Fig. 9.17. Crane operations.

linear differential equation of the form

$$M \frac{d^2\eta_3}{dt^2} + B \frac{d\eta_3}{dt} + C\eta_3 = F(t) \tag{9.47}$$

with initial conditions $\eta_3 = 0$ and $d\eta_3/dt = 0$. Use the fact that the solution of this equation system can be written as

$$\eta_3(t) = \int_0^t h(t - \tau)F(\tau)\,d\tau \tag{9.48}$$

where

$$h(t) = \frac{1}{M\omega_0} e^{-\lambda\omega_0 t} \sin \omega_0 t,$$

$\omega_0 = (C/M)^{\frac{1}{2}}$ and $\lambda = B/(2M\omega_0) \ll 1$. Set M equal to the mass of the cylinder, $B = 0$ and C equal to the stiffness of the crane-wire system. Use equation (9.40) to calculate the slamming. Neglect the influence of the body velocity on the slamming force.

Calculate numerically the dynamic tension in the wire during a period $2\pi/\omega_0$ just after the cylinder hits the water. Compare the maximum wire tension during the impact period with the tension in the wire before impact. Compare the results with a quasi-static solution where the impact force at $t = 0$ is applied as a static force.

9.4   Wave loads on bow-flare sections

Equations (9.37) can be generalized and applied for a ship at forward speed. An example is hydrodynamic forces on bow-flare sections. The hydrodynamic analyses will be based on strip theory. In this reference frame the forward speed of the ship appears as an incident steady flow with velocity U along the positive x-axis (see Fig. 9.18). We will start with the momentum equation (9.29). The control surfaces S can be selected as shown in Fig. 9.18. There are two lateral planes S_0 separated by a distance dx exterior to the ship. The intersection curve between the planes and the ship is called $\Sigma_B(x)$. We will limit ourselves to consider forced heave motion and no incident waves.

(a) Show that the vertical force on the ship cross-section can be written as

$$F_3 = -\rho \frac{d}{dt} \iint\limits_{S_B+S_F} \phi n_3\,ds - \iint\limits_{S_B+S_F} \rho gz n_3\,ds$$

$$- \rho \iint\limits_{S_\infty+S_0} [V_3 V_n - \tfrac{1}{2}|\mathbf{V}|^2 n_3]\,ds \tag{9.49}$$

(Positive normal direction is out of the fluid domain).

(b) Show that the integral over S_0 in equation (9.49) can be written as

$$-\rho U \, dx \frac{\partial}{\partial x} \int_{\Sigma_B(x)} \phi n_3 \, ds \qquad (9.50)$$

(c) Show that the vertical hydrodynamic force $F_3^{(2D)}$ per unit length can be written as

$$F_3^{(2D)} = -\left(\frac{d}{dt} - U \frac{\partial}{\partial x}\right)\left(A_{33}^{(2D)}(x) \frac{d\eta_3}{dt}\right) + \rho g A(x) \qquad (9.51)$$

where $A(x)$ is the submerged cross-sectional area.

(d) How would you propose to generalize the formula by including the effect of incident waves and pitch motion. (Hint: See equations (9.19) and (9.43)).

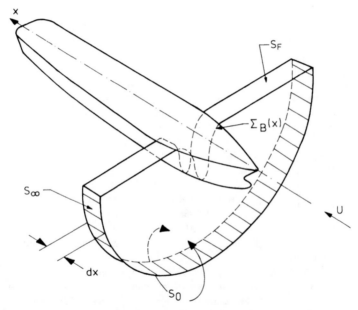

Fig. 9.18. Use of momentum equation to calculate wave loads on bow-flare sections of ships (see exercise 9.4.)

REFERENCES

Aarsnes, J. V. (1984). Current forces on ships. Dr. ing. thesis, Report UR-84-39, Dept. of Marine Technology, Nor. Inst. Technol, Trondheim.

Aarsnes, J. V., Faltinsen, O. M. & Pettersen, B. (1985). Application of a vortex tracking method to current forces on ships. *Proc. Conf. Separated Flow Around Marine Structures*, pp. 309–46, Trondheim: Nor. Inst. Technol.

Abramowitz, M. & Stegun, I. A. (1964). *Handbook of Mathematical Functions with Formulas, Graphs and Mathematical Tables*, 1046 pp., New York: Dover Publications Inc.

Abramson, H. N., Bass, R. L., Faltinsen, O. M. & Olsen, H. A. (1974). Liquid Slosh in LNG Carriers. In *Proc. Tenth Symp. on Naval Hydrodynamics*, ed. R. D. Copper & S. W. Doroff, pp. 371–88, Arlington, Va.: Office of Naval Research – Department of the Navy.

Achenbach, E. (1968). Distribution of local pressure around a circular cylinder in cross flow up to $Re = 5 \cdot 10^6$. *J. Fluid Mech.*, **34**, 625–39.

Achenbach, E. (1971). Influence of surface roughness on the cross-flow around a circular cylinder. *J. Fluid Mech.*, **46**, 321–35.

Achenbach, E. & Heinecke, E. (1981). On vortex shedding from smooth and rough cylinders in the range of Reynolds numbers $8 \cdot 10^3$ to $5 \cdot 10^8$, *J. Fluid Mech.*, **109**, 239–51.

Aquirre, J. E. & Boyce, T. R. (1974). Estimation of wind forces on offshore drilling platforms. *Trans. Royal Inst. Nav. Arch.* (RINA), **116**, 93–109.

Armand, J. L. & Cointe, R. (1986). Hydrodynamic impact analysis of a cylinder. In *Proc. Fifth Int. Offshore Mech. and Arctic Engng. Symp.* (OMAE), Tokyo, Japan, vol. 1, pp. 609–34. New York: The American Society of Mechanical Engineering.

Bai, K. J. & Yeung, R. W. (1974). Numerical solutions to free surface problems. In *Proc. Tenth Symp. on Naval Hydrodynamics*, ed. R. D. Cooper & S. W. Doroff, pp. 609–34, Arlington, Va.: Office of Naval Research – Department of the Navy.

Batchelor, G. K. (1970). *An Introduction to Fluid Dynamics*. Cambridge: Cambridge University Press.

Bearman, P. W. (1985). Vortex trajectories in oscillatory flow. In *Proc. Separated Flow around Marine Structures*, pp. 133–54, Trondheim: Nor. Inst. Technol.

Bearman, P. W. (1988). Wave loading experiments on circular cylinders at large scale. In *Proc. Int. Conf. Behaviour of Offshore Struct.*, (BOSS '88), ed. T. Moan, N. Janbu & O. Faltinsen, vol. 2, pp. 471–87. Trondheim: Tapir Publishers.

Bearman, P. W., Graham, J. M. R. & Sing, S. (1979). Forces on cylinders in harmonically oscillating flow. In *Proc. Mechanics of Wave-Induced Forces on Cylinders*, ed. T. L. Shaw, pp. 437–49, London: Pitman.

Bearman, P. W., Graham, J. M. R., Naylor, P. & Obasaju, E. D. (1981). The role of vortices in oscillatory flow about bluff cylinders. In *Proc. Int. Symp. on Hydrodynamics in Ocean Engng*, vol. 2, pp. 621–35. Trondheim: Nor. Inst. Technol.

Bearman, P. W., Graham, J. M. R., Obasaju, E. D. & Drossopoulos, G. H. (1984a). The influence of corner radius on the forces experienced by cylindrical bluff bodies in oscillatory flow. *Appl. Ocean Res.*, **6**, 2, 83–9.

Bearman, P. W., Graham, J. M. R., Obasaju, E. D. (1984b). A model equation for the transverse force on cylinders in oscillatory flows. *Appl. Ocean Res.*, **6**, 3, 166–72.

Blevins, R. D. (1977). *Flow-Induced Vibration*. New York: Van Nostrand Reinhold Company.

Brown, C. E. & Michael, W. H. (1955). On slender delta wings with leading edge separation. *Nat. Adv. Comm. Aero Tech.* Note 3430.

Børresen, R. (1984). The unified theory of ship motions in water of finite depth. Dr. ing. thesis, Div. Marine Hydrodynamics, Nor. Inst. Technol., Trondheim.

Campbell, I. M. C. & Weynberg, P. A. (1980). Measurement of parameters affecting slamming. Final Report, Rep. No. 440, *Technology Reports Centre No. OT-R-8042*. Southampton University: Wolfson Unit for Marine Technology.

Chaplin, J. R. (1984). Nonlinear forces on a horizontal cylinder beneath waves, *J. Fluid Mech.*, **147**, 449–64.

Chaplin, J. R. (1988). Nonlinear forces on horizontal cylinders in the Inertia Regime in Waves at High Reynolds numbers. In *Proc. Int. Conf. Behaviour of Offshore Struct.*, (BOSS '88), ed. T. Moan, N. Janbu & O. Faltinsen, vol. 2, pp. 505–18. Trondheim: Tapir Publishers.

Chislett, M. S. & Bjørheden, O. (1966). *Influence of Ship Speed on the Effectiveness of a Lateral-Thrust-Unit*. Report No. Hy-8, Lyngby: Hydro- og Aerodynamisk Laboratorium.

Chorin, A. J. (1973), Numerical study of slightly viscous flow, *J. Fluid Mech.*, **57**, 785–96.

Davenport, A. G. (1978). The prediction of the response of structure to gusty wind. In *Safety of Structures under Dynamic Loading*, ed. I. Holand, D. Kavlie, G. Moe & R. Sigbjørnsen, vol. 1, pp. 257–84. Trondheim: Tapir Publishers.

Delany, N. K. & Sorensen, N. E. (1953). *Low-speed Drag of Cylinders of Various Shapes*. Washington: NACA Technical Note 3038.

Dommermuth, D. G., Yue, D. K. P., Lin, W. M., Rapp, R. J., Chan, E. S. & Melville, W. K. (1988). Deep water plunging breakers: A comparison between potential theory and experiments, *J. Fluid Mech.*, **189**, 423–42.

Fage, A. & Johansen, F. C. (1928). The structure of vortex sheets. *Aeronautical Research Council R and M 1143* also *Phil. Mag.*, **5**, 114–42.

Fage, A. & Warsap, J. H., (1929). *The Effects of Turbulence and Surface Roughness on the Drag of Circular Cylinders*. A.R.C. Rep. and Memo No. 1283.

Faltinsen, O. M. (1972). Wave forces on a restrained ship in head-sea waves. In *Proc. Ninth Symp. on Naval Hydrodynamics*, ed. R. Brard and A. Castera, vol.

2, pp. 1763–844. Arlington, Va.: Office of Naval Research – Department of the Navy.

Faltinsen, O. M. & Michelsen, F. (1974). Motions of large structures in waves at zero Froude number. In *Int. Symp. Dynamics of Marine Vehicles and Structures in Waves*, ed. R. E. D. Bishop & W. G. Price, pp. 91–106. London: Mechanical Engineering Publications Ltd.

Faltinsen, O. M. & Pettersen, B. (1987). Application of a vortex tracking method to separated flow around marine structures. *J. Fluids Struct.*, **1**, 217–37.

Faltinsen, O. M. & Sortland, B. (1987). Slowdrift eddy making damping of a ship. *Appl. Ocean Res.*, **9**, 1, 37–46.

Faltinsen, O. M. & Zhao, R. (1989). Slow-drift motions of a moored two-dimensional body in irregular waves. *J. Ship Res.*, **33**, 2, 93–106.

Faltinsen, O. M., Kjaerland, O., Liapis, N. & Walderhaug, H. (1979). Hydrodynamic analysis of tankers at single-point mooring systems. In *Proc. 2nd Int. Conf. Behaviour of Offshore Struct.*, (BOSS '79), ed. H. S. Stephans & S. M. Knight, vol. 2, pp. 177–206. Cranfield, Bedford: BHRA Fluid Engineering.

Faltinsen, O. M., Minsaas, K., Liapis, N. & Skjørdal, S. O. (1980). Prediction of resistance and propulsion of a ship in a seaway. In *Proc. Thirteenth Symp. on Naval Hydrodynamics*, ed. T. Inui, pp. 505–30. Tokyo: The Shipbuilding Research Association of Japan.

Faltinsen, O. M., van Hooff, R. W., Fylling, I. & Teigen, P. (1982). Theoretical and experimental investigations of tension leg platform (TLP) behaviour. In *Proc. Behaviour of Offshore Struct.*, (BOSS '82), ed. C. Chryssostomidis & J. J. Connor, vol. 1, pp. 411–23. Washington: Hemisphere Publishing Corporation.

Faltinsen, O. M., Dahle, L. A. & Sortland, B. (1986). Slowdrift damping and response of moored ship in irregular waves. In *Proc. Fifth Int. Offshore Mech. and Arctic Engng. Symp.*, (OMAE), vol. 1, pp. 297–303. New York: The American Society of Mechanical Engineers.

Feng, C. C. (1968). The measurement of vortex-induced effects in flow past stationary and oscillating circular and D-section cylinders, M.Sc. Thesis, U. British Columbia.

Frank, W. (1967). *Oscillation of Cylinders in or Below the Free Surface of Deep Fluids*, Report 2375. Washington DC: Naval Ship Research and Development Center.

Gerritsma, J. & Beukelman, W. (1972). Analysis of the resistance increase in waves of a fast cargo ship. *Intern Shipbuilding Progr.*, **19**, 217, 285–93.

Gerritsma, J., Beukelman, W. & Glansdorp, C. C. (1974). The effects of beam on the hydrodynamic characteristics of ship hulls. In *Proc. Tenth Symp. on Naval Hydrodynamics*, ed. R. D. Cooper & S. W. Doroff, pp. 3–34. Arlington, Va.: Office of Naval Research – Department of the Navy.

Graham, J. M. R. (1980). The forces on sharp-edged cylinder in oscillatory flow at low Keulegan–Carpenter numbers. *J. Fluid Mech.*, **97**, 331–46.

Greenhow, M. (1988). Water-entry and -exit of a horizontal circular cylinder. *Appl. Ocean Res.*, **10**, 4, 191–8.

Greenhow, M. & Lin, W.-M. (1983). *Non-linear free surface effects: experiments and theory*. Report No. 83-19, Dept. Ocean Engng. Cambridge, Mass: Mass. Inst. Technol.

Greenhow, M. & Li Yanbao (1987). Added mass for circular cylinders near to or penetrating fluid boundaries – review, extension and application to water-entry, -exit and slamming. *Ocean Engng.*, **14**, 4, 325–48.

Griffin, O. M. (1985). The effect of current shear on vortex shedding. In *Proc. Conf. Separated Flow Around Marine Structures*, pp. 91–110. Trondheim: Nor. Inst. Technol.

Griffin, O. M., Skop, R. A. & Ramberg, S. E. (1975). The resonant vortex excited vibrations of structures and cable systems. In *Proc. Offshore Technol. Conf.* (OTC), paper no. 2319, vol. 2, pp. 731–44. Houston: Offshore Technology Conference.

Grue, J. & Palm, E. (1985). Wave radiation and wave diffraction from a submerged body in a uniform current. *J. Fluid Mech.*, **151**, 257–78.

Hagiwara, K. & Yuhara, T. (1976). Fundamental study of wave impact loads on ship bow. Selected papers from the *J. Soc. Nav. Arch.*, Japan, **14**, 73–85.

Havelock, T. H. (1942). The damping of the heaving and pitching motion of a ship. *Phil. Mag.*, **33**, 7, 666–73.

Havelock, T. H. (1955). Waves due to floating sphere making periodic heaving oscillations. *Proc. Roy. Soc. Lond.*, **A231**, 1–7.

Hess, J. L. & Smith, A. M. O. (1962). Calculation of non-lifting potential flow about arbitrary three-dimensional bodies. Report No. E.S. 40622, Douglas Aircraft Division, Long Beach, California. (See also *J. Ship Res.*, (1964), **8**, 2, 22–44.)

Hoerner, S. F. (1965). *Fluid Dynamic Drag*. Published by the author.

Hughes, G. (1954). Friction and form resistance in turbulent flow, and a proposed formulation for use in model and ship correlation. *Transaction of the Institution of Naval Architects*, **96**, 314–76.

Huse, E. (1986). Influence of mooring line damping upon rig motions. In *Proc. 18th Offshore Technol. Conf.* (OTC), paper no. 5204, vol. 2, pp. 433–38, Houston: Offshore Technology Conference.

Ikeda, Y., Himeno, Y. & Tanaka, N., (1976). On roll damping force of ship: Effects of friction of hull and normal force of bilge keels. *J. Kansai Society of Nav. Arch.*, Japan, **161**, 41–51.

Ikeda, Y., Himeno, Y. & Tanaka, N. (1977a). On eddy making component of roll damping force on naked hull. *J. Soc. Nav. Arch.*, Japan, **142**, 54–66.

Ikeda, Y., Komatsu, K., Himeno, Y. & Tanaka, N. (1977b). On roll damping force of ship. Effect of hull surface pressure created by bilge keels. *J. Kansai Soc. of Nav. Arch.*, Japan, **165**, 31–40.

Isherwood, R. M., (1973). Wind resistance of merchant ships. *Trans. Inst. Naval Arch.*, (RINA), **115**, 327–38.

John, F. (1950). On the motion of floating bodies II. *Communications on Pure and Applied Mathematics*, **3**, 45–101.

Johnson, I. G. (1978). A new approach to oscillatory rough turbulent boundary layers. Series Paper 17, Institute of Hydrodynamic and Hydraulic Engineering, Technical University of Denmark, Lyngby.

Kato, H. (1966). Effect of bilge keels on the rolling of ships. *Mem. of the Defence Academy*, Japan, **IV**, 3, 369–84.

Keulegan, G. H. & Carpenter, L. H. (1958). Forces on cylinders and plates in an oscillating fluid. *J. Res. National Bureau of Standards*, **60**, 5, 423–40.

Kim, M. H. & Yue, D. K. P. (1988). The non-linear sum-frequency wave excitation and response of a tension-leg platform. In *Proc. Int. Conf. Behaviour of Offshore Struct.*, (BOSS '88), ed. T. Moan, N. Janbu, & O. Faltinsen, vol. 2, pp. 687–704, Trondheim: Nor. Inst. Technol.

Kinsman, B. (1965). *Wind Waves*. Englewood Cliffs, N.J.: Prentice–Hall Inc.

Koehler, B. R. & Kettleborough, C. F. (1977). Hydrodynamic impact of a falling body upon a viscous incompressible fluid. *J. Ship. Res.*, **21**, 3, 165–81.

Korsmeyer, F. T., Lee, C. H., Newman, J. N. & Sclavounos, P. D. (1988). The analysis of wave effects on tension leg platforms. In *Proc. Seventh Int. Conf. Offshore Mech. and Arctic Engng.*, (OMAE), vol. 2, pp. 1–14, New York: The American Society of Mechanical Engineering.

Lamb, H. (1945). *Hydrodynamics*. New York: Dover Publications.

Lecointe, Y. & Piquet, J. (1985). Compact finite-difference methods for solving incompressible Navier-Stokes equations around oscillatory bodies. Von Karman Institute for Fluid Dynamics Lecture Series 1985-04, Computational Fluid Dynamics.

Lee, C.-H. & Sclavounos, P. D. (1989). The removal of irregular frequencies from integral equations in wave–body interactions. *J. Fluid Mech.* **207**, 393–418.

Lee, W. T., Bales, W. L. & Sowby, S. E. (1985). Standardized wind and wave environments for North Pacific Ocean Areas. R/SPD-0919-02, DTNSRDC, Washington D.C.

Liapis, N. (1985). Wave loads and motion stability of tankers at single point mooring systems. Dr. ing. Thesis Rep. No. UR-85-44. Dept. of Marine Technology, Nor. Inst. Technol., Trondheim.

Longuet-Higgins, M. S. (1953). Can sea waves cause microseisms? In *Proc. Symp. on Microseisms*, publ. no. 306, pp. 74–93. New York: U.S. Nat. Sc. – Nat. Res. Council.

Longuet-Higgins, M. S. (1977). The mean forces exerted by waves on floating or submerged bodies with applications to sand bars and wave power machines. *Proc. Roy. Soc. Lond.*, **A352**, 463–80.

Lundgren, H. (1969). Wave shock forces: An analysis of deformations and forces in the wave and in the foundations. *Proc. Symp. Res. on Wave Action*, vol. 2, paper no. 4, 20 pp. Delft: Delft Hydraulics Labs.

Maruo, H. (1960). The drift of a body floating in waves. *J. Ship. Res.*, **4**, 3, 1–10.

McCamy, R. S. & Fuchs, R. A. (1954). Wave forces on piles: A diffraction theory. No. 69, Beach Erosion Board, US Army Corps of Engineers, Techn. Memo, Washington.

Mei, C. C. (1983). *The Applied Dynamics of Ocean Surface Waves*. New York: John Wiley & Sons. Revised printing (1989), Singapore: World Scientific.

Miloh, T. (1981). Wave slam on a sphere penetrating a free-surface. *J. Eng. Math.*, **15**, 3, 221–40.

Minsaas, K. J., Thon, H. J. & Kauczynski, W. (1986). Influence of Ocean Environment on Thruster Performance. In *Proc. Int. Symp. Propeller and Cavitation*, supplementary volume, pp. 124–42. Shanghai: The Editorial Office of Shipbuilding of China.

Morison, J. R., O'Brien, M. P., Johnson, J. W. & Schaaf, S. A. (1950). The force exerted by surface waves on piles. *Pet. Trans.*, **189**, 149–54.

Newman, J. N. (1962). The exciting forces on fixed bodies in waves. *J. Ship Res.*, **6**, 4, 10–17.

Newman, J. N. (1967). The drift force and moment on ships in waves. *J. Ship Res.*, **11**, 1, 51–60.

Newman, J. N. (1974). Second order, slowly varying forces on vessels in irregular waves. In *Proc. Int. Symp. Dynamics of Marine Vehicles and Structures in Waves*, ed. R. E. D. Bishop & W. G. Price, pp. 182–6. London: Mechanical Engineering Publications Ltd.

Newman, J. N. (1977). *Marine Hydrodynamics*. Cambridge: The MIT Press.

Newman, J. N. (1985). Algorithms for the free-surface Green function, *J. Eng. Math.*, **19**, 57–67.

Newman, J. N. (1990). Second-harmonic wave diffraction at large depths. *J. Fluid Mech.* **213**, 59–70.

Newman, J. N. & Sclavounos, P. D. (1988). The computation of wave loads on large offshore structures. In *Proc. Int. Behaviour of Offshore Struct.*, (BOSS '88), eds T. Moan, N. Janbu & O. Faltinsen, vol. 2, pp. 605–22. Trondheim: Tapir Publishers.

NORDFORSK (1987). The Nordic Cooperative project. Seakeeping performance of Ships, *Assessment of a ship performance in a seaway*, 1987, Trondheim, Norway: MARINTEK.

Ochi, M. K. (1964). Prediction of occurrence and severity of ship slamming at sea. In *Proc. Fifth Symp. on Naval Hydrodynamics*, pp. 545–96. Washington DC: Office of Naval Research – Department of the Navy.

Ochi, M. K. (1982). Stochastic analysis and probability prediction in random seas. *Advances in Hydroscience*, **13**, 217–375.

Ochi, M. K. & Motter, L. E. (1974). Prediction of extreme ship responses in rough seas of the North Atlantic. In *Proc. Int. Symp. Dynamics of Marine Vehicles and Structures in Waves*, ed. R. E. D. Bishop & W. G. Price, pp. 187–97. London: Mechanical Engineering Publications Ltd.

Ogilvie, T. F. (1963). First- and second-order forces on a cylinder submerged under a free surface. *J. Fluid Mech.*, **16**, 451–72.

Ogilvie, T. F. (1964). Recent progress towards the understanding and prediction of ship motions. In *Proc. Fifth Symp. on Naval Hydrodynamics*, pp. 3–128. Washington DC: Office of Naval Research – Department of the Navy.

Ogilvie, T. F. (1983). Second order hydrodynamic effects on ocean platforms. In *Proc. Int. Workshop on Ship and Platform Motions*, ed. R. W. Yeung, pp. 205–65. Berkeley: Continuing Education in Eng., University extension, University of California, Berkeley.

Ohkusu, M. & Iwashita, H. (1987). Radiation and diffraction wave pattern of ships with forward speed. *Trans. West-Japan Soc. Nav. Arch.*, **73**, 1–18.

Overvik, T. (1982). Hydroelastic motion of multiple risers in a steady current. Dr. ing. Thesis, Div. Port and Ocean Engng. Nor. Inst. Technol., Trondheim.

Parkinson, G. V. (1985). Hydroelastic phenomena of bodies of bluff section in steady flow. In *Proc. Conf. Separated Flow around Marine Structures*, pp. 37–58. Trondheim: Nor. Inst. Technol.

Pinkster, J. A. (1975). Low-frequency phenomena associated with vessels moored at sea. *Society of Petroleum Engineers Journal*, **December,** 487–94.

Pinkster, J. A. & van Oortmerssen, G. (1977). Computation of the first- and second-order wave forces on oscillating bodies in regular waves. In *Proc.*

Second Int. Conf. Numerical Ship Hydrodynamics, ed. J. V. Wehausen & N. Salvesen, pp. 136–56, Berkeley: University Extension Publications, University of California, Berkeley.

Price, W. G. & Bishop, R. E. D. (1974). *Probabilistic Theory of Ship Dynamics*. London: Chapman and Hall Ltd.

Roshko, A. (1961). Experiments on the flow past a circular cylinder at very high Reynolds number. *J. Fluid Mech.*, **10**, 345–56.

Routh, E. J. (1955). *Advanced Dynamics of a System of Rigid Bodies*. New York: Dover Publication.

Salvesen, N., Tuck, E. O. & Faltinsen, O. M. (1970). Ship motions and sea loads. *Trans. SNAME*, **78**, 250–87.

Sarpkaya, T. (1978). Fluid forces on oscillating cylinders. *J. Waterway*, Port, Coastal and Ocean Division of ASCE, **104**, 275–90.

Sarpkaya, T. (1985). Past progress and outstanding problems in time-dependent flows about ocean structures. In *Proc. Conf. Separated Flow around Marine Structures*, pp. 1–36, Trondheim: Nor. Inst. Technol.

Sarpkaya, T. (1986). Force on a circular cylinder in viscous oscillatory flow at low Keulegan–Carpenter numbers. *J. Fluid Mech.*, **165**, 61–71.

Sarpkaya, T. & Isaacson, M., (1981). *Mechanics of Wave Forces on Offshore Structures*. New York: Van Nostrand Reinhold Company.

Sarpkaya, T. & Shoaff, R. L. (1979). A discrete vortex analysis of flow about stationary and transversely oscillating cylinders. Tech. Rep. NPS-69 SL 79011, *Nav. Postgrad. Sch.*, Monterey, California.

Schlichting, H. (1979). *Boundary-Layer Theory*. New York: McGraw-Hill Book Company.

Schmidt, D. V. & Tilman, P. M. O. (1972). On the development of the circulation in water of circular cylinders. *Acustica*, **27**, 14–22.

Schmitke, R. T. & Murdey, D. C. (1980). Seakeeping and Resistance Trade-Offs in frigate hull form designs. In *Proc. Thirteenth Symp. on Naval Hydrodynamics*, ed. T. Inui, pp. 455–78. Tokyo: The Shipbuilding Research Association of Japan.

Schwartz, L. W. (1974). Computer extension and analytic continuation of Stokes' expansion for gravity waves. *J. Fluid Mech.*, **62**, 553–78.

Skop, R. A., Griffin, O. M. & Ramberg, S. E. (1977). Streaming predictions for the Seacon II Experimental Mooring., In *Proc. Offshore Technol. Conf.* (OTC), paper no. 2884, vol. 3, pp. 61–6. Houston: Offshore Technology Conference.

Smith, P. A. & Stansby, P. K. (1988). Impulsively started flow around a circular cylinder by the vortex method. *J. Fluid. Mech.*, **194**, 45–77.

Sortland, B. (1986). Force measurements in oscillatory flow on ship sections and circular cylinders in a U-tube water tank. Dr. ing. Thesis, Report No. UR-86-52. Dept. of Marine Technology, Nor. Inst. Technol., Trondheim.

Sparks, C. P. (1980). Mechanical behaviour of marine risers mode of influence of principal parameters. *Trans. ASME*, **102**, 214–22.

Stokes, G. G. (1851). *Trans. Camb. Phil. Soc.* 9.8, Mathematical and Physical Papers 3,1.

Strom-Tejsen, J., Yen, H. Y. H. & Moran D. D. (1973). Added resistance in waves. *Trans. SNAME*, **81**, 109–43.

Tanaka, N. (1961). A study on the bilge keels. (Part 4. On the eddy making resistance to the rolling of a ship hull). *J. Soc. Nav. Arch.*, Japan, **109**, 205–12.

Tanaka, N., Ikeda, Y. & Nishino, K. (1982). Hydrodynamic viscous force acting on oscillating cylinders with various shapes. In *Proc. 6th Symp. of Marine Technology*, The Society of Naval Architects of Japan. (Also *Rep. Dep. Nav. Arch.*, University of Osaka Prefecture, no. 407, Jan. 1983).

Tasai, F. (1959). On the damping force and added mass of ships heaving and pitching, vol. VII, no. 26, *Rep. of Res. Inst. for Appl. Mech.*, Kyushu University.

Teliónis, D. P. (1981). *Unsteady Viscous Flows*, New York: Springer–Verlag New York Inc.

Telste, J. G. (1987). Inviscid flow about a cylinder rising to a free-surface. *J. Fluid Mech.*, **182**, 149–68.

Verhagen, J. H. G. (1967). The impact of a flat plate on a water surface. *J. Ship. Res.*, **11**, 4, 211–23.

Verley, R. L. P. (1982). A simple model of vortex-induced forces in waves and oscillating currents. *Appl. Ocean Res.*, **4**, 2, 117–20.

Vinje, T. (1980). Statistical distributions of hydrodynamic forces on objects in current and waves. *Norwegian Maritime Research*, **8**, 2, 20–6.

von Karman, T. (1929). The impact of seaplane floats during landing. NACA, Technical Note 321, Washington.

Vugts, J. M. (1968). The hydrodynamic coefficients for swaying, heaving and rolling cylinders in a free surface. Report No. 112 S, Neth. Ship. Ship Res. Centre T.N.O.

Wagner, H. (1932). Über Stoss- und Gleitvorgänge an der Oberfläche von Flüssigkeiten. *Zeitschr. f. Angewandte Mathematik und Mechanik*, **12**, 4, 193–235.

Wang, C.-Y., (1968). On high-frequency oscillating viscous flows. *J. Fluid Mech.*, **32**, 55–68.

Wehausen, J. V. & Laitone, E. V. (1960). Surface waves. In *Handbuch der Physik*, vol. 9, pp. 446–778, Berlin: Springer–Verlag.

Wichers, J. E. W. (1982). On the low-frequency surge motions of vessels moored in high seas. In *Proc. 14th Offshore Technol. Conf.*, (OTC), paper no. 4437, vol. 4, pp. 711–34, Houston: Offshore Technology Conference.

Williamson, C. H. K. (1985). Sinusoidal flow relative to circular cylinders. *J. Fluid Mech.*, **155**, 141–74.

Yeung, R. (1982). Numerical methods in free-surface flows. *Annu. Rev. Fluid Mech.*, **14**, 395–442.

Zdravkovich, M. M. (1985). Forces on pipe clusters. In *Proc. Conf. Separated Flow around Marine Structures*, pp. 201–26, Trondheim: Nor. Inst. Technol.

Zhao, R. & Faltinsen, O. M. (1988). A comparative study of theoretical models for slowdrift sway motions of a marine structure. In *Proc. Seventh Int. Conf. Offshore Mech. and Arctic Engng.*, (OMAE), vol. 2, pp. 153–58. New York: The American Society of Mechanical Engineering.

Zhao, R., Faltinsen, O. M., Krokstad, J. R. & Aanesland, V. (1988). Wave-current interaction effects on large-volume structures. In *Proc. Int. Conf. Behaviour of Offshore Structures*, (BOSS '88), ed. T. Moan, N. Janbu & O. Faltinsen, vol. 2, pp. 623–38, Trondheim: Tapir Publishers.

INDEX